Physical Methods in Chemistry and Nano Science.
Volume 1: Elemental Analysis

Physical Methods in Chemistry and Nano Science.
Volume 1: Elemental Analysis

Editor

Andrew R. Barron

Contributors

Andrew R. Barron, Simon G. Bott, Brandon Cisneros,
Julian Cooper, Laurie Cullum, Johnathan Dietz,
Huilong Fei, Natalia Gonzalez Pech, Amanda M. Goodman,
Lauren Harrison, Jessica Heimann, Meghan Jebb, Lei Li,
Danielle Michaud, Alvin Orbaek White, Graham Piburn,
Hannah Rutledge, Avishek Saha

MiDAS Green Innovations
2020

MiDAS Green Innovation, Ltd
Swansea, SA1 8RD, UK

www.midasgreeninnovation.com

Dedication

To all the undergraduate and graduate students who have had to suffer my lectures over the last 33 years.

"Everybody in the whole of the world, feels the same inside"

Marillion (1989)

Contents

Acknowledgements

I would like to thank all the contributors to this Volume, for their interest in creating a user-friendly text for their peers.

The myriad collaborators of whom I have had the pleasure of learning from over the years are to be thanked for bringing new characterization methods to my research - you know who you are.

Last, but not least, I would like to thank my wife, Merrie, for putting up with me during the Editing of this book while 'social distancing' during the COVID-19 pandemic.

Preface

This Series intended as a survey of research techniques used in modern chemistry, materials science, and nanoscience. The topics are grouped into volumes, not be method *per se*, but with regard to the type of information that can be obtained. Thus, the Volumes are ordered as follows:

- Elemental composition.
- Physical and thermal analysis.
- Chromatography
- Chemical speciation.
- Molecular and solid state structure.
- Surface morphology and structure at the nanoscale.
- Device performance.
- Applications of analytical methods

As a consequence of this organization methods can be found in different Volumes. For example, X-ray photoelectron spectroscopy is included under Elemental Composition (Volume 1) with regard to its use for determining the chemical composition, while it is included under Chemical Speciation (Volume 4) with regard to determining the identity of component chemical moieties.

The goal was to create simple to understand explanations of methods that allow the reader to gain the knowledge to correctly apply a technique or interpret data. As a consequence, the topics in this book have been developed in partnership with undergraduate and postgraduate students at Rice University over a 7-year period, and because of this there is some variation in depth and focus given to each topic. I make no apology for this diversity.

Chapter 1: Introduction

Andrew R. Barron

The purpose of elemental analysis is to determine the quantity of a particular element within a molecule or material. Elemental analysis can be subdivided in two ways:

- Qualitative: determining what elements are present or the presence of a particular element.
- Quantitative: determining how much of a particular or each element is present.

In either case elemental analysis is independent of structure unit or functional group, i.e., the determination of carbon content in toluene ($C_6H_5CH_3$) does not differentiate between the aromatic sp^2 carbon atoms and the methyl sp^3 carbon, but rather defines the overall chemical formula, i.e., C_7H_8.

Elemental analysis can be performed on a solid, liquid, or gas. However, depending on the technique employed the sample may have to be pre-reacted, e.g., by combustion or acid digestion. The amounts required for elemental analysis range from a few gram (g) to a few milligram (mg) or less.

Elemental analysis can also be subdivided into general categories related to the approach involved in determining quantities.

- Classical analysis relies on stoichiometry through a chemical reaction or by comparison with known reference sample.
- Modern methods rely on nuclear structure or size (mass) of a particular element and are generally limited to solid samples.

Many classical methods they can be further classified into the following categories:

- Gravimetric in which a sample is separated from solution as a solid as a precipitate and weighed. This is generally used for alloys, ceramics, and minerals.
- Volumetric is the most frequently employed involves determination of the volume of a substance that combines with another substance in known proportions. This is also called titrimetric analysis and is frequently employed using a visual end point or potentiometric measurement.

- Colorimetric (spectroscopic) analysis requires the addition of an organic complex agent. This is commonly used in medical laboratories as well as in the analysis of industrial wastewater treatment.

The biggest limitation in classical methods is mainly due to sample manipulation rather than equipment error, i.e., operator error in weighing a sample or observing an end point. In contrast, the errors in modern analytical methods are almost entirely computer sourced and inherent in the software that analyzes and fits the data.

Chapter 2: Spot Tests

Andrew R. Barron

Spot tests (spot analysis) are simple chemical procedures that uniquely identify a substance. They can be performed on small samples, even microscopic samples of matter with no preliminary separation. The first report of a spot test was in 1859 by Hugo Shi for the detection of uric acid (Figure 2.1); however, several spot tests are specific for elements rather than compounds.

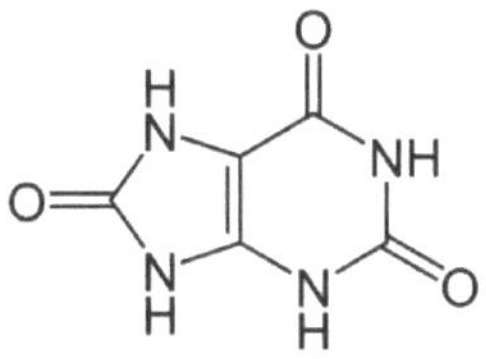

Figure 2.1: The structure of uric acid

In a typical spot test, a drop of chemical reagent is added to a drop of an unknown mixture. If the substance under study is present, it produces a chemical reaction characterized by one or more unique observables, e.g., a color change.

Detection of chlorine

A typical example of a spot test is the detection of chlorine in the gas phase by the exposure to paper impregnated with 0.1% 4-4'-*bis*-dimethylamino-thiobenzophenone (thio-Michler's ketone) dissolved in benzene. In the presence of chlorine, the paper will change from yellow to blue. The mechanism involves the zwitter ionic form of the thioketone,

undergoing an oxidation reaction and subsequent disulfide coupling,

$$2 \; Me_2N\text{—}C(S^{\cdot})\text{=}N^+Me_2 + 2\,Cl_2 \longrightarrow (Me_2N\text{—}C\text{=}N^+Me_2)_2(S\text{—}S) + 2\,Cl^-$$

Bibliography

L. Ben-Dor and E. Jungreis, A direct selective spot test for chlorine in inorganic and organic compounds. *Microchimica Acta*, 1964, **52**, 100.

F. Feigl, *Spot Tests in Organic Analysis*, 7th edn., Elsevier, New York (2012).

A. N. MacInnes, A. R. Barron, J. J. Li, and T. R. Gilbert, Reaction bonded refractory metal carbide/carbon composites. *Polyhedron*, 1994, **13**, 1315

N. MacInnes, A. R. Barron, R. S. Soman, and T. R. Gilbert, Reaction-bonded niobium carbide on graphite via a novel solution impregnation process. *J. Am. Ceram. Soc.*, 1990, **73**, 3696.

C. L. O'Neal, D. J. Crouch, and A. A. Fatah, Validation of twelve chemical spot tests for the detection of drugs of abuse. *Forensic Sci. Int.*, 2000, **109**, 189.

M. Tubino, A. V. Rossi, and M. E. A. de Magalhães, Quantitative spot tests of Fe(III), Cr(VI) and Ni(II) by reflectance measurements. *Anal. Lett.*, 1997, **30**, 271.

Chapter 3: Combustion Analysis

Laurie Cullum and Andrew R. Barron

Applications of combustion analysis

Combustion, or burning as it is more commonly known, is simply the mixing and exothermic reaction of a fuel and an oxidizer. It has been used since pre-historic times in a variety of ways, such as a source of direct heat, as in furnaces, boilers, stoves, and metal forming, or in piston engines, gas turbines, jet engines, rocket engines, guns, and explosives. Automobile engines use internal combustion in order to convert chemical energy into mechanical energy. Combustion is utilized in the production of large quantities of H_2. Coal or coke is combusted at 1000 °C in the presence of water in a two-step reaction. The first step involved the partial oxidation of carbon to carbon monoxide:

$$C(g) + H_2O(g) \rightarrow CO(g) + H_2(g)$$

The second step involves a mixture of produced carbon monoxide with water to produce hydrogen and is commonly known as the water gas shift reaction:

$$CO(g) + H_2O(g) \rightarrow CO_2(g) + H_2(g)$$

Although combustion provides a multitude of uses, it was not employed as a scientific analytical tool until the late 18th century.

History of combustion

In the 1780's, Antoine Lavoisier (Figure 3.1) was the first to analyze organic compounds with combustion using an extremely large, and at the time, expensive apparatus (Figure 3.2) that required a team of operators and over 50 g of the organic sample being analyzed.

The method was simplified and optimized throughout the 19th and 20th centuries, first by Joseph Gay-Lussac (Figure 3.3), who began to use copper oxide in 1815, which is still used as the standard oxidation catalyst.

Figure 3.1: French chemist and renowned "father of modern chemistry" Antoine Lavoisier (1743 - 1794).

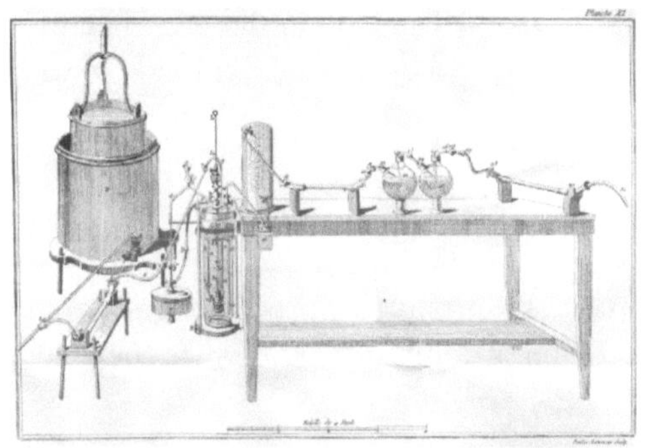

Figure 3.2: Lavoisier's combustion apparatus. A. Lavoisier, *Traité Élémentaire de Chimie, présenté dans un ordre nouveau, et d'après des découvertes modernes*, Cuchet, Paris (1789).

Figure 3.3: French chemist Joseph Gay-Lussac (1778 - 1850).

William Prout (Figure 3.4) invented a new method of combustion analysis in 1827 by heating a mixture of the sample and CuO using a multiple-flame alcohol lamp (Figure 3.5) and measuring the change in gaseous volume.

Figure 3.4: English chemist, physician, and natural theologian William Prout (1785 - 1850).

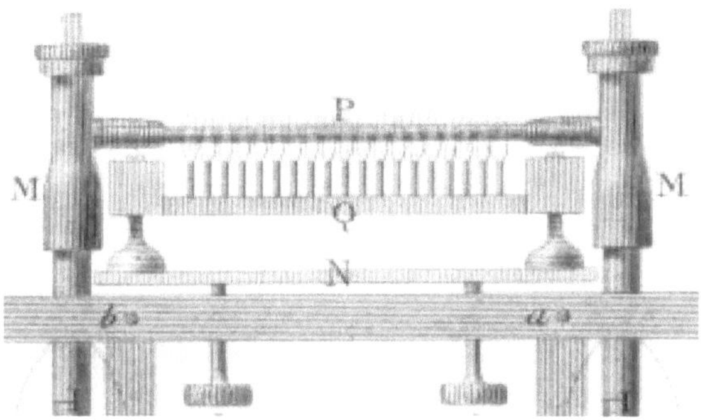

Figure 3.5: Prout's combustion apparatus. Adapted from W. Prout, On the ultimate composition of simple alimentary substances; with some preliminary remarks on the analysis of organized bodies in general. *Philos. T. R. Soc. Lond.*, 1827, 117, 355.

In 1831, Justus von Liebig (Figure 3.6) simplified the method of combustion analysis into a *combustion train* system (Figure 3.7 and Figure 3.8) that linearly heated the sample using coal, absorbed water using calcium chloride ($CaCl_2$), and absorbed carbon dioxide using potash (KOH). This new method only required 0.5 g of sample and a single operator. Liebig moved the sample through the apparatus by sucking on an opening at the far-right end of the apparatus (Figure 3.7).

Figure 3.6: German chemist Justus von Liebig (1803 - 1873).

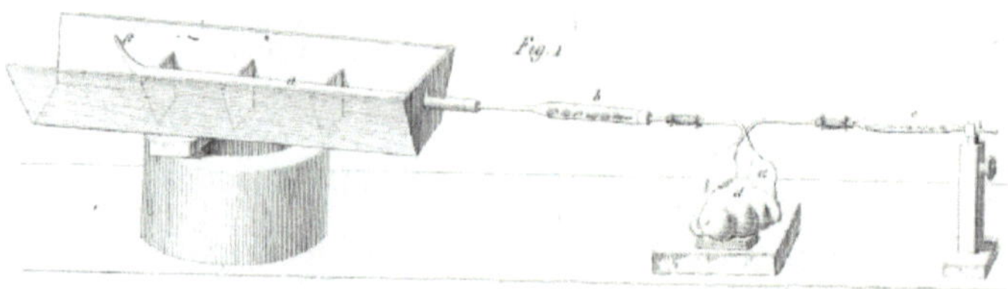

Figure 3.7: Print of von Liebig's combustion train apparatus for determining carbon and hydrogen composition. J. von Liebig, Vermischte chemische No-tizen. *Annalen der Physik und Chemie*, 1831, 21, 1.

Figure 3.8: Photograph of von Liebig's combustion train apparatus for determining carbon and hydrogen composition. The Oesper Collections in the History of Chemistry, Apparatus Museum, University of Cincinnati, Case 10, Combustion Analysis. For a 360° view of this apparatus, visit http://digitalprojects.libraries.uc.edu/oesper/museum/case10/shelf_02/CA0010/index.php.

Jean-Baptiste André Dumas (Figure 3.9) used a similar combustion train to Liebig; however, he added a U-shaped aspirator that prevented atmospheric moisture from entering the apparatus (Figure 3.10).

Figure 3.9: French chemist Jean-Baptiste André Dumas (1800 - 1844).

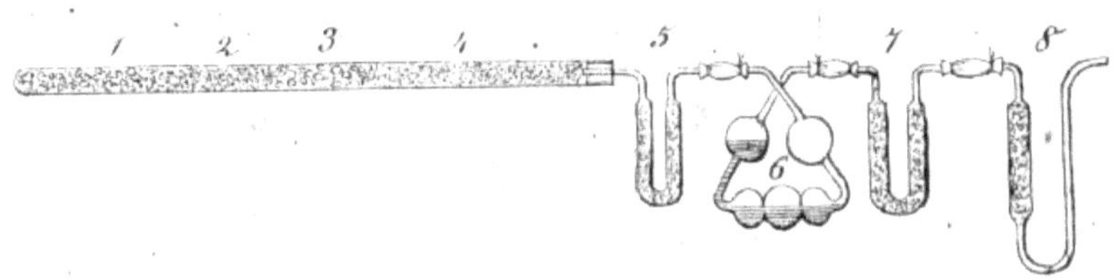

Figure 3.10: Dumas' apparatus; note the aspirator at 8. Adapted from J. A. Dumas, *Ann. der Chem. and Pharm.*, 1841, 38, 141.

In 1923, Fritz Pregl (Figure 3.11) received the Nobel Prize for inventing a micro-analysis method of combustion analysis. This method required only 5 mg or less, which is 0.01% of the amount required in Lavoisier's apparatus. Today, combustion analysis of an organic or organometallic compound only requires about 2 mg of sample.

Although this method of analysis destroys the sample and is not as sensitive as other techniques, it is still considered a necessity for characterizing an organic compound.

Figure 3.11: Austrian chemist and physician Fritz Pregl (1869 - 1930).

Categories of combustion

Basic flame types

There are several categories of combustion, which can be identified by their flame types (Table 3.1). At some point in the combustion process, the fuel and oxidant must be mixed together. If these are mixed before being burned, the flame type is referred to as a premixed flame, and if they are mixed simultaneously with combustion, it is referred to as a non-premixed flame. In addition, the flow of the flame can be categorized as either laminar (streamlined) or turbulent (Figure 3.12).

Fuel/oxidizer mixing	Fluid motion	Examples
Premixed	Turbulent	Spark-ignited gasoline engine, low NO_x stationary gas turbine
Premixed	Laminar	Flat flame, Bunsen flame (followed by a non-premixed candle for $\Phi>1$)
Non-premixed	Turbulent	Pulverized coal combustion, aircraft turbine, diesel engine, H_2/O_2 rocket motor
Non-premixed	Laminar	Wood fire, radiant burners for heating, candle

Table 1.1: Types of combustion systems with examples. Adapted from J. Warnatz, U. Maas, and R. W. Dibble, *Combustion: Physical and Chemical Fundamentals, Modeling and Simulation, Experiments, Pollutant Formation*, 3rd Ed., Springer, Berlin (2001).

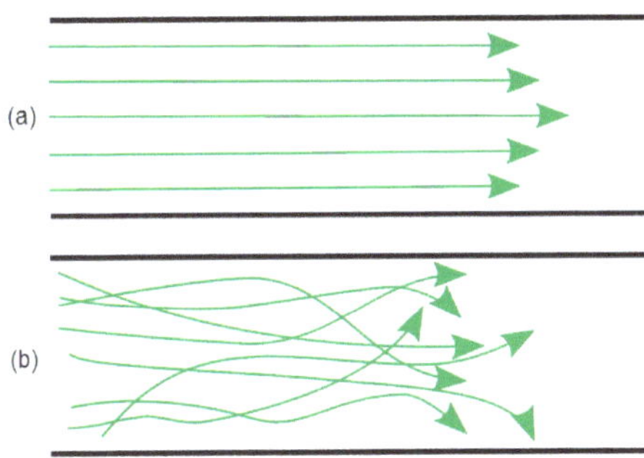

Figure 3.12: Schematic representation of (a) laminar flow and (b) turbulent flow.

The amount of oxygen in the combustion system can alter the flow of the flame and the appearance. As illustrated in Figure 3.13, a flame with no oxygen tends to have a very turbulent flow, while a flame with an excess of oxygen tends to have a laminar flow.

Figure 3.13: Bunsen burner flames with varying amounts of oxygen and constant amount of fuel. (1) air valve completely closed, (2) air valve slightly open, (3) air valve half open, (4) air valve completely open.

Stoichiometric combustion and calculations

A combustion system is referred to as stoichiometric when all of the fuel and oxidizer are consumed, and only carbon dioxide and water are formed. On the other hand, a fuel-rich system has an excess of fuel, and a fuel-lean system has an excess of oxygen (Table 3.2).

Combustion type	Reaction example
Stoichiometric	$2H_2 + O_2 \rightarrow 2H_2O$
Fuel-rich (H_2 left over)	$3H_2 + O_2 \rightarrow 2H_2O + H_2$
Fuel-lean (O_2 left over)	$CH_4 + 3O_2 \rightarrow 2H_2O + CO_2 + O_2$

Table 3.2: Examples of stoichiometric, fuel-rich, and fuel-lean systems.

If the reaction of a stoichiometric mixture is written to describe the reaction of exactly 1 mol of fuel (H_2 in this case), then the mole fraction of the fuel content can be easily calculated as follows, where v denotes the mole number of O_2 in the combustion reaction equation for a complete reaction to H_2O and CO_2:

$$x_{fuel,stoich.} = \frac{1}{1 + v}$$

For example, in the reaction,

$$H_2 + \tfrac{1}{2}O_2 \rightarrow H_2O_2 + H_2$$

the stoichiometry is determined as shown in,

$$v = 0.5$$

$$x_{H_2,stoich.} = 2/3$$

However, as calculated this reaction would be for the reaction in an environment of pure oxygen. On the other hand, air has only 21% oxygen (78% nitrogen, 1% noble gases). Therefore, if air is used as the oxidizer, this must be taken into account in the calculations,

$$x_{N_2} = 3.762(x_{O_2})$$

The mole fractions for a stoichiometric mixture in air are therefore calculated in following way:

$$x_{fuel,stoich} = \frac{1}{1 + v(4.762)}$$

$$x_{O_2,stoich} = \nu(x_{fuel,stoich})$$

$$x_{N_2,stoich} = 3.762\,(x_{O_2,stoich})$$

For example, calculation of the fuel mole fraction (x_{fuel}) for the stoichiometric reaction:

$$CH_4 + 2O_2 + (2 \times 3.762)N_2 \rightarrow CO_2 + 2H_2O + (2 \times 3.762)N_2$$

In this reaction $\nu = 2$, as 2 moles of oxygen are needed to fully oxidize methane into H_2O and CO_2,

$$x_{fuel,stoich} = \frac{1}{1 + (2 \times 4.762)} = 0.09502 = 9.502\ mol\%$$

Premixed combustion reactions can also be characterized by the air equivalence ratio, λ, as shown in,

$$\Phi = 1/\lambda$$

The fuel equivalence ratio, Φ, is the reciprocal of this value,

$$\lambda = \frac{x_{air}/x_{fuel}}{x_{air,stoich}/x_{fuel,stoich}}$$

Rewriting in terms of the fuel equivalence ratio gives:

$$x_{fuel} = \frac{1}{1 + (\nu\,4.672/\Phi)}$$

$$x_{air} = 1 - x_{fuel}$$

$$x_{O_2} = x_{air}/4.762$$

$$x_{N_2,stoich} = 3.762(x_{O_2})$$

The premixed combustion processes can also be identified by their air and fuel equivalence ratios (Table 3.3).

Type of combustion	Φ	λ
Rich	>1	<1
Stoichiometric	=1	=1
Lean	<1	>1

Table 3.3: Identification of combustion type by Φ and λ values.

With a premixed type of combustion, there is much greater control over the reaction. If performed at lean conditions, then high temperatures, the pollutant nitric oxide, and the production of soot can be minimized or even avoided, allowing the system to combust efficiently. However, a premixed system requires large volumes of premixed reactants, which pose a fire hazard. As a result, non-premixed combusted, while not being efficient, is more commonly used.

Instrumentation

Though the instrumentation of combustion analysis has greatly improved, the basic components of the apparatus (Figure 3.14) have not changed much since the late 18th century.

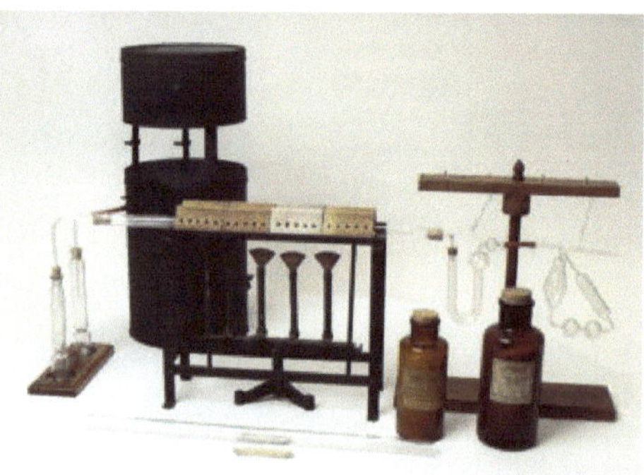

Figure 3.14: Combustion apparatus from the 19th century. The Oesper Collections in the History of Chemistry, Apparatus Museum, University of Cincinnati, Case 10, Combustion Analysis. For a 360° view of this apparatus, visit http://digitalprojects.libraries.uc.edu/oesper/museum/case10/shelf_03/CA0012/index.php.

The sample of an organic compound, such as a hydrocarbon, is contained within a furnace or exposed to a flame and burned in the presence of oxygen, creating water vapor and carbon dioxide gas (Figure 3.15). The sample moves first through the apparatus to a chamber in which H_2O is absorbed by a hydrophilic substance and second through a chamber in which CO_2 is absorbed. The change in weight of each chamber is determined to calculate the weight of H_2O and CO_2. After the masses of H_2O and CO_2 have been determined, they can be used to characterize and calculate the composition of the original sample.

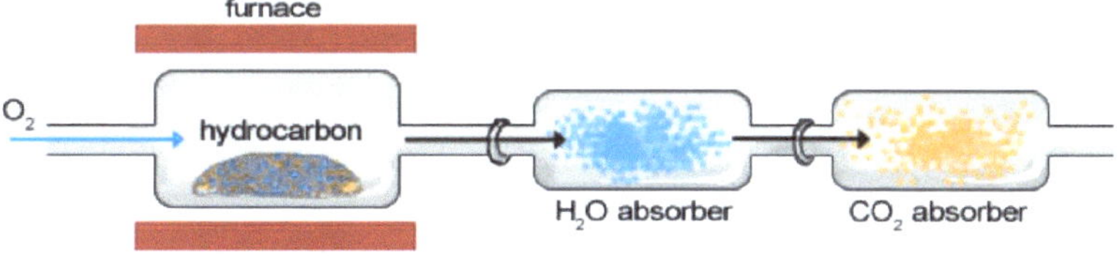

Figure 3.15: Typical modern combustion apparatus with a furnace.

Calculations and determining chemical formulas

Hydrocarbons

Combustion analysis is a standard method of determining a chemical formula of a substance that contains hydrogen and carbon. First, a sample is weighed and then burned in a furnace in the presence of excess oxygen. All of the carbon is converted to carbon dioxide, and the hydrogen is converted to water in this way. Each of these are absorbed in separate compartments, which are weighed before and after the reaction. From these measurements, the chemical formula can be determined.

Generally, the following reaction takes place in combustion analysis:

$$C_aH_b + O_2\,(xs) \rightarrow aCO_2 + {}^b/_2\,H_2O$$

For example, after burning 1.333 g of an unknown hydrocarbon (C_xH_y) in a combustion analysis apparatus, 1.410 g of H_2O and 4.305 g of CO_2 were produced. Separately, the molar mass of this hydrocarbon was found to be 204.35 g/mol. The following represents the method employed to calculate the empirical and molecular formulas of the unknown hydrocarbon.

Step 1: Using the molar masses of water and carbon dioxide, determine the moles of hydrogen and carbon that were produced.

$$1.410\text{g } H_2O \ \times \ \frac{1 \text{ mol } H_2O}{18.015 \text{ g } H_2O} \ \times \ \frac{2 \text{ mol } H}{1 \text{ mol } H_2O} \ = \ 0.1565 \text{ mol } H$$

$$4.3051 \text{ g } CO_2 \ \times \ \frac{1 \text{ mol } CO_2}{44.010 \text{ g } CO_2} \ \times \ \frac{1 \text{ mol } C}{1 \text{ mol } CO_2} \ = \ 0.09782 \text{ mol } C$$

Step 2: Divide the larger molar amount by the smaller molar amount. In some cases, the ratio is not made up of two integers. Convert the numerator of the ratio to an improper fraction and rewrite the ratio in whole numbers as shown.

$$\frac{0.1565 \text{ mol } H}{0.09782 \text{ mol } C} \ = \ \frac{1.600 \text{ mol } H}{1 \text{ mol } C} \ = \ \frac{13/5 \text{ mol } H}{1 \text{ mol } C} \ = \ \frac{8/5 \text{ mol } H}{1 \text{ mol } C} \ = \ \frac{8 \text{ mol } H}{5 \text{ mol } C}$$

Therefore, the empirical formula of the unknown hydrocarbon is C_5H_8.

Step 3: To get the molecular formula, divide the experimental molar mass of the unknown hydrocarbon by the empirical formula weight.

$$\frac{\text{Molar mass}}{\text{Empirical formula weight}} \ = \ \frac{204.35 \text{ g/mol}}{68.114 \text{ g/mol}} \ = \ 3$$

Therefore, the molecular formula of the hydrocarbon is $(C_5H_8)_3$ or $C_{15}H_{24}$. Some of the multitude of possible compounds with the chemical formula of $C_{15}H_{24}$ are shown in Figure 3.16. It should be noted that without additional characterization, it is impossible to further define the identity of the unknown hydrocarbon.

Compounds containing carbon, hydrogen, and oxygen

Combustion analysis can also be utilized to determine the empiric and molecular formulas of compounds containing carbon, hydrogen, and oxygen. However, as the reaction is performed in an environment of excess oxygen, the amount of oxygen in the sample can be determined from the sample mass, rather than the combustion data.

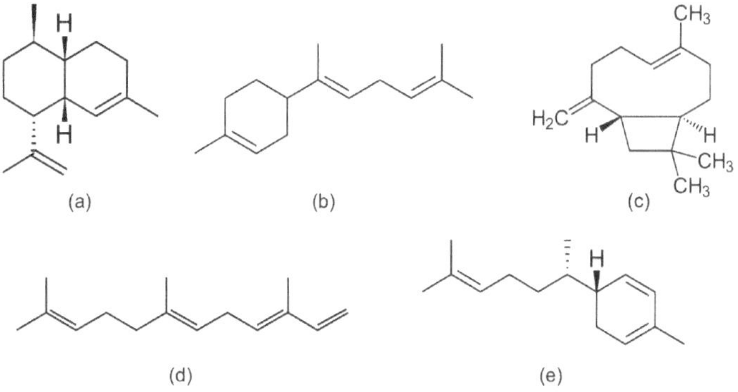

Figure 3.16: Selected possible compounds with the molecular formula $C_{15}H_{24}$: (a) amorpha-4,11-diene, (b) α-bisabolene, (c) caryophyllene, (d) (*E,E*)-α-farnesene, and (e) zingiberene

For example, a 2.0714 g sample containing carbon, hydrogen, and oxygen was burned in a combustion analysis apparatus; 1.928 g of H_2O and 4.709 g of CO_2 were produced. Separately, the molar mass of the sample was found to be 116.16 g/mol. Determine the empirical formula, molecular formula, and identity of the sample.

Step 1: Using the molar masses of water and carbon dioxide, determine the moles of hydrogen and carbon that were produced.

$$1.928 \text{ g } H_2O \ \times \ \frac{1 \text{ mol } H_2O}{18.015 \text{ g } H_2O} \ \times \ \frac{2 \text{ mol } H}{1 \text{ mol } H_2O} \ = \ 0.2140 \text{ mol } H$$

$$4.709 \text{ g } CO_2 \ \times \ \frac{1 \text{ mol } CO_2}{44.010 \text{ g } CO_2} \ \times \ \frac{1 \text{ mol } C}{1 \text{ mol } CO_2} \ = \ 0.1070 \text{ mol } C$$

Step 2: Using the molar amounts of carbon and hydrogen, calculate the masses of each in the original sample.

$$0.2140 \text{ mol } H \ \times \ \frac{1.008 \text{ g } H}{1 \text{ mol } H} \ = \ 0.2157 \text{ g } H$$

$$0.1070 \text{ mol C} \times \frac{12.011 \text{ g C}}{1 \text{ mol C}} = 1.285 \text{ g C}$$

Step 3: Subtract the masses of carbon and hydrogen from the sample mass. Now that the mass of oxygen is known, use this to calculate the molar amount of oxygen in the sample.

$$2.0714 \text{ g sample} - 0.2157 \text{ g H} - 1.285 \text{ g C} = 0.5707 \text{ g O}$$

$$0.5707 \text{ g O} \times \frac{1 \text{ mol O}}{16.00 \text{ g O}} = 0.03567 \text{ mol O}$$

Step 4: Divide each molar amount by the smallest molar amount in order to determine the ratio between the three elements.

$$\frac{0.03567 \text{ mol O}}{0.03567} = 1.00 \text{ mol O} = 1 \text{ mol O}$$

$$\frac{0.1070 \text{ mol C}}{0.03567} = 3.00 \text{ mol C} = 3 \text{ mol C}$$

$$\frac{0.2140 \text{ mol H}}{0.03567} = 5.999 \text{ mol H} = 6 \text{ mol H}$$

Therefore, the empirical formula is C_3H_6O.

Step 5: To get the molecular formula, divide the experimental molar mass of the unknown hydrocarbon by the empirical formula weight.

$$\frac{\text{Molar mass}}{\text{Empirical formula weight}} = \frac{116.16 \text{ g/mol}}{58.08 \text{ g/mol}} = 2$$

Therefore, the molecular formula is $(C_3H_6O)_2$ or $C_6H_{12}O_2$. Possible compounds with this molecular formula are shown in (Figure 3.17).

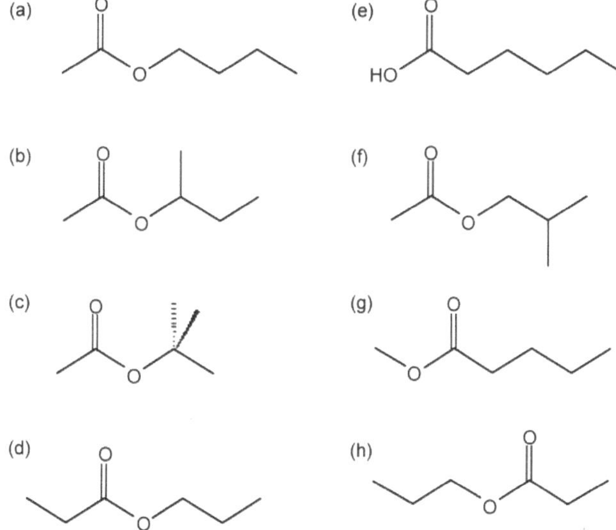

Figure 3.17: Structures of possible compounds with the molecular formula $C_6H_{12}O_2$: (a) butylacetate, (b) *sec*-butyl acetate, (c) *tert*-butyl acetate, (d) ethyl butyrate, (e) haxanoic acid, (f) isobutyl acetate, (g) methyl pentanoate, and (h) propyl proponoate.

Binary compounds

By using combustion analysis, the chemical formula of a binary compound containing oxygen can also be determined. This is particularly helpful in the case of combustion of a metal which can result in potential oxides of multiple oxidation states.

For example, a sample of iron weighing 1.7480 g is combusted in the presence of excess oxygen. A metal oxide (Fe_xO_y) is formed with a mass of 2.4982 g. Determine the chemical formula of the oxide product and the oxidation state of Fe.

Step 1: Subtract the mass of Fe from the mass of the oxide to determine the mass of oxygen in the product.

2.4982 g Fe_xO_y - 1.7480 g Fe = 0.7502 g O

Step 2: Using the molar masses of Fe and O, calculate the molar amounts of each element.

$$1.7480 \text{ g Fe} \times \frac{1 \text{ mol Fe}}{55.845 \text{ g Fe}} = 0.031301 \text{ mol Fe}$$

$$0.7502 \text{ g O} \times \frac{1 \text{ mol O}}{16.00 \text{ g O}} = 0.04689 \text{ mol O}$$

Step 3: Divide the larger molar amount by the smaller molar amount. In some cases, the ratio is not made up of two integers. Convert the numerator of the ratio to an improper fraction and rewrite the ratio in whole numbers as shown.

$$\frac{0.031301 \text{ mol Fe}}{0.04689 \text{ mol O}} = \frac{0.6675 \text{ mol Fe}}{1 \text{ mol O}} = \frac{^2/_3 \text{ mol Fe}}{1 \text{ mol O}} = \frac{2 \text{ mol Fe}}{3 \text{ mol O}}$$

Therefore, the chemical formula of the oxide is Fe_2O_3, and Fe is in the 3+ oxidation state.

Bibliography

J. A. Dumas, *Ann. Chem. Pharm.*, 1841, **38**, 141.

H. Goldwhite, Gay-Lussac after 200 years. *J. Chem. Edu.*, 1978, **55**, 366.

A. Lavoisier, *Traité Élémentaire de Chimie, présenté dans un ordre nouveau, et d'après des découvertes modernes*, Cuchet, Paris (1789).

A. Linan and F. A. Williams, *Fundamental Aspects of Combustion*, Oxford University Press, New York (1993).

W. Prout, On the ultimate composition of simple alimentary substances; with some preliminary remarks on the analysis of organized bodies in general. *Philos. T. R. Soc. Lond.*, 1827, **117**, 355.

D. Shriver and P. Atkins, *Inorganic Chemistry*, 5th edn., W. H. Freeman and Co., New York (2009).

J. von Liebig, Vermischte chemische Notizen. *Annalen der Physik und Chemie*, 1831, **21**, 1.

J. Warnatz, U. Maas, and R. W. Dibble, *Combustion: Physical and Chemical Fundamentals, Modeling and Simulation, Experiments, Pollutant Formation*, 3rd edn., Springer, Berlin (2001).

Chapter 4: Atomic Absorption Spectroscopy

Hannah Rutledge and Andrew R. Barron

History

The earliest spectroscopy was first described by Marcus Marci von Kronland (Figure 4.1) in 1648 by analyzing sunlight as is passed through water droplets and thus creating a rainbow. Further analysis of sunlight by William Hyde Wollaston (Figure 4.2) led to the discovery of black lines in the spectrum, which in 1820 Sir David Brewster (Figure 4.3) explained as absorption of light in the sun's atmosphere.

Figure 4.1: Bohemian physicist Johannes Marcus Marci von Kronland (1595 - 1667).

Robert Bunsen (Figure 4.4) and Gustav Kirchho (Figure 4.5) both studied the sodium spectrum and came to the conclusion that every element has its own unique spectrum that can be used to identify elements in the vapor phase. Kircho further explained the phenomenon by stating that if a material can emit radiation of a certain wavelength, that it may also absorb radiation of that wavelength. Although Bunsen and Kircho took a large step in defining the technique of atomic absorption spectroscopy (AAS), it was not widely utilized as an analytical technique, except in the field of astronomy, for many years due to many practical difficulties.

Figure 4.2: English chemist and physicist William Hyde Wollaston (1659 - 1724).

Figure 4.3: Scottish physicist, mathematician, astronomer, inventor, writer and university principal Sir David Brewster (1781 - 1868).

Figure 4.4: German chemist Robert Bunsen (1811 - 1899).

Figure 4.5: German physicist Gustav Robert Kirchho (1824 - 1887).

In 1953, Alan Walsh (Figure 4.6) drastically improved the AAS methods. He advocated AAS to many instrument manufacturers, but to no avail. Although he had improved the methods, he hadn't shown how it could be useful in any applications. In 1957, he discovered uses for AAS that convinced manufactures market the first commercial AAS spectrometers. Since that time, AAS's popularity has fluctuated as other analytical techniques and improvements to the methods are made.

Figure 4.6: British physicist Sir Alan Walsh (1916 - 1988).

Theory of atomic absorption spectroscopy

In order to understand how atomic absorption spectroscopy works, some background information is necessary. Our modern view of atomic theory began with John Dalton (Figure 4.7) in the 18th century when he proposed the concept of atoms, that all atoms of an element are identical, and that atoms of different elements can combine to form molecules.

Figure 4.7: English chemist, physicist, and meteorologist John Dalton FRS (1766 - 1844).

In 1913, Niels Bohr (Figure 4.8) revolutionized atomic theory by proposing quantum numbers, a positively charged nucleus, and electrons orbiting around the nucleus in the what became known as the Bohr model of the atom. Soon afterward, Louis deBroglie (Figure 4.9) proposed quantized energy of electrons, which is an extremely important concept in AAS. Wolfgang Pauli (Figure 4.10) then elaborated on deBroglie's theory by stating that no two electrons can share the same four quantum numbers. These landmark discoveries in atomic theory are necessary in understanding the mechanism of AAS.

Figure 4.8: Danish physicist Niels Henrik David Bohr (1885 - 1962).

Figure 4.9: French physicist and a Nobel laureate Louis de Broglie (1892 - 1987).

Figure 4.10: Austrian physicist Wolfgang Pauli (1900 - 1958).

Atoms have valence electrons, which are the outermost electrons of the atom. Atoms can be excited when irradiated, which creates an absorption spectrum. When an atom is excited, the valence electron moves up an energy level. The energies of the various stationary states, or restricted orbits, can then be determined by these emission lines. The resonance line is then defined as the specific radiation absorbed to reach the excited state.

The Maxwell-Boltzmann equation gives the number of electrons in any given orbital. It relates the distribution to the thermal temperature of the system (as opposed to electronic temperature, vibrational temperature, or rotational temperature). Plank proposed radiation emitted energy in discrete packets (quanta),

$$E = h\nu$$

which can be related to Einstein's equation,

$$E = mc^2$$

Both atomic emission and atomic absorption spectroscopy can be used to analyze samples. Atomic emission spectroscopy measures the intensity of light emitted by the excited atoms, while atomic absorption spectroscopy measures the light absorbed by atomic absorption. This light is typically in the visible or ultraviolet region of the electromagnetic spectrum. The percentage is then compared to a calibration curve to determine the amount of material in the sample. The energy of the system can be used to find the frequency of the radiation, and thus the wavelength through:

$$\nu = c/\lambda$$

Because the energy levels are quantized, only certain wavelengths are allowed, and each atom has a unique spectrum. There are many variables that can affect the system. For example, if the sample is changed in a way that increases the population of atoms, there will be an increase in both emission and absorption and vice versa. There are also variables that affect the ratio of excited to unexcited atoms such as an increase in temperature of the vapor.

Applications of atomic absorption spectroscopy

There are many applications of atomic absorption spectroscopy (AAS) due to its specificity. These can be divided into the broad categories of biological analysis, environmental and marine analysis, and geological analysis.

Biological analysis

Biological samples can include both human tissue samples and food samples. In human tissue samples, AAS can be used to determine the amount of various levels of metals and other electrolytes, within tissue samples. These tissue samples can be many things including but not limited to blood, bone marrow, urine, hair, and nails. Sample preparation is dependent upon the sample. This is extremely important in that many elements are toxic in certain concentrations in the body, and AAS can analyze what concentrations they are present in. Some examples of trace elements that samples are analyzed for are arsenic, mercury, and lead.

An example of an application of AAS to human tissue is the measurement of the electrolytes, sodium and potassium in plasma. This measurement is important because the values can be indicative of various diseases when outside of the normal range. The typical method used for this analysis is atomization of a 1:50 dilution in strontium chloride ($SrCl_2$) using an air-hydrogen flame. The sodium is detected at its secondary line (330.2 nm) because detection at the first line would require further dilution of the sample due to signal intensity. The reason that strontium chloride is used is because it reduces ionization of the potassium and sodium ions, while eliminating phosphate's and calcium's interference.

In the food industry, AAS provides analysis of vegetables, animal products, and animal feeds. These kinds of analyses are some of the oldest application of AAS. An important consideration that needs to be taken into account in food analysis is sampling. The sample should be an accurate representation of what is being analyzed. Because of this, it must be homogenous, and many it is often needed that several samples are run. Food samples are most often run in order to determine mineral and trace element amounts so that consumers know if they are consuming an adequate amount. Samples are also analyzed to determine heavy metals which can be detrimental to consumers.

Environmental and marine analysis

Environmental and marine analysis typically refers to water analysis of various types. Water analysis includes many things ranging from drinking water to wastewater to sea water. Unlike biological samples, the preparation of water samples is governed more by laws than by the sample itself. The analytes that can be measured also vary greatly and can often include lead, copper, nickel, and mercury.

An example of water analysis is an analysis of leaching of lead and zinc from tin-lead solder into water. The solder is what binds the joints of copper pipes. As a comparison, soft water, acidic water, and chlorinated water all analyzed. The sample preparation consisted of exposing the various water samples to copper plates with solder for various intervals of time. The samples are then analyzed for copper and zinc with air-acetylene flame AAS using a deuterium lamp. If the samples have a copper level below 100 µg/L, the method was changed to graphite furnace electrothermal AAS due to its higher sensitivity.

Geological analysis

Geological analysis encompasses both mineral reserves and environmental research. When prospecting mineral reserves, the method of AAS used needs to be cheap, fast, and versatile because the majority of prospects end up being of no economic use. When studying rocks, preparation can include acid digestions or leaching. If the sample needs to have silicon content analyzed, acid digestion is not a suitable preparation method.

An example is the analysis of lake and river sediment for lead and cadmium. Because this experiment involves a solid sample, more preparation is needed than for the other examples. The sediment was first dried, then grounded into a powder, and then was decomposed in a bomb with nitric acid (HNO_3) and perchloric acid ($HClO_4$). Standards of lead and cadmium were prepared. Ammonium sulfate ($[NH_4][SO_4]$) and ammonium phosphate ($[NH_4][H_3PO_4]$) were added to the samples to correct for the interferences caused by sodium and potassium that are present in the sample. The standards and samples were then analyzed with electrothermal AAS.

Instrumentation

Atomizer

In order for the sample to be analyzed, it must first be atomized. This is an extremely important step in AAS because it determines the sensitivity of the reading. The most effective atomizers create a large number of homogenous free atoms. There are many types of atomizers, but only two are commonly used:

- flame atomizer,
- electrothermal atomizer.

Flame atomizer

Flame atomizers Figure 4.11 are widely used for a multitude of reasons including their simplicity, low cost, and long length of time that they have been utilized. Flame atomizers accept an aerosol from a nebulizer into a flame that has enough energy to both volatilize and atomize the sample (Figure 4.11). When this happens, the sample is dried, vaporized, atomized, and ionized. Within this category of atomizers, there are many subcategories determined by the chemical composition of the flame. The composition of the flame is often determined based on the sample being analyzed. The flame itself should meet several requirements including sufficient energy, a long length, non-turbulent, and safe.

Electrothermal atomizer

Although electrothermal atomizers were developed before flame atomizers, they did not become popular until more recently due to improvements made to the detection level. They employ graphite tubes that increase temperature in a stepwise manner (Figure 4.12). Electrothermal atomization first dries the sample and evaporates much of the solvent and impurities, then atomizes the sample, and then rises it to an extremely high temperature to clean the graphite tube. Some requirements for this form of atomization are the ability to maintain a constant temperature during atomization, have rapid atomization, hold a large volume of solution, and emit minimal radiation. Electrothermal atomization is much less harsh than the method of flame atomization.

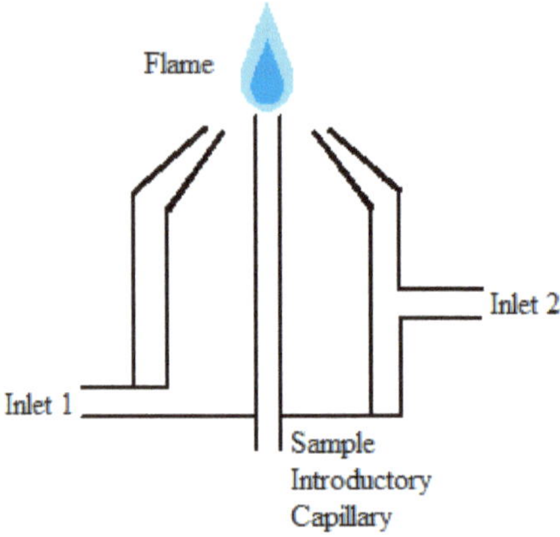

Figure 4.11: A schematic diagram of a flame atomizer showing the oxidizer inlet (1) and fuel inlet (2).

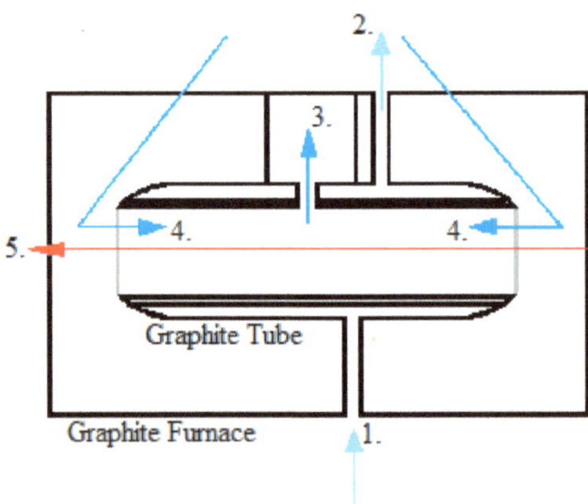

Figure 4.12: Schematic diagram of an electrothermal atomizer showing the external gas ow inlet (1), the external gas ow outlet (2), the internal gas ow outlet (3), the internal gas flow inlet (4), and the light beam (5).

Radiation source

The radiation source then irradiates the atomized sample. The sample absorbs some of the radiation, and the rest passes through the spectrometer to a detector. Radiation sources can be separated into two broad categories: line sources and continuum sources.

Line sources excite the analyte and thus emit its own line spectrum. Hollow cathode lamps and electrodeless discharge lamps are the most commonly used examples of line sources. On the other hand, continuum sources have radiation that spreads out over a wider range of wavelengths. These sources are typically only used for background correction. Deuterium lamps and halogen lamps are often used for this purpose.

Spectrometer

Spectrometers are used to separate the different wavelengths of light before they pass to the detector. The spectrometer used in AAS can be either single-beam or double-beam. Single-beam spectrometers only require radiation that passes directly through the atomized sample, while double-beam spectrometers Figure 4.13, as implied by the name, require two beams of light; one that passes directly through the sample, and one that does not pass through the sample at all. The single-beam spectrometers have less optical components and therefore suffer less radiation loss. Double-beam monochromators have more optical components, but they are also more stable over time because they can compensate for changes more readily.

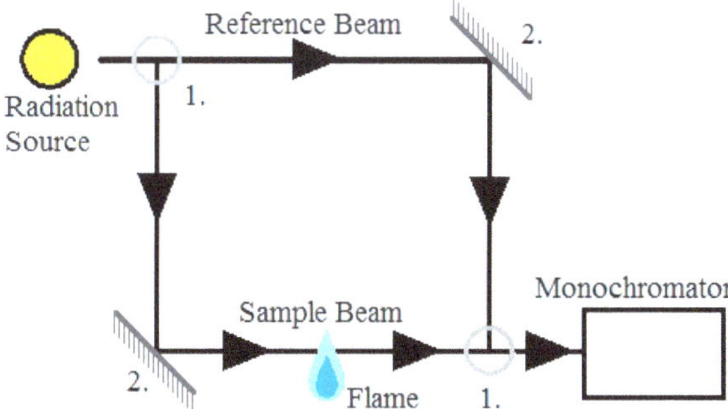

Figure 4.13: A schematic of a double-beam spectrometer showing the 50/50 beam splitters (1) and the mirrors (2).

Obtaining measurements

Sample preparation

Sample preparation is extremely varied because of the range of samples that can be analyzed. Regardless of the type of sample, certain considerations should be made. These include: the laboratory environment, the vessel holding the sample, storage of the sample, and pretreatment of the sample.

Sample preparation begins with having a clean environment to work in. AAS is often used to measure trace elements, in which case contamination can lead to severe error. Possible equipment includes laminar flow hoods, clean rooms, and closed, clean vessels for transportation of the sample. Not only must the sample be kept clean, it also needs to be conserved in terms of pH, constituents, and any other properties that could alter the contents.

When trace elements are stored in a container, the vessel walls can adsorb some of the analyte leading to poor results. To correct for this, perfluoroalkoxy polymers (PFA), silica, glassy carbon, and other materials with inert surfaces are often used as the storage material. Acidifying the solution with hydrochloric or nitric acid can also help prevent ions from adhering to the walls of the vessel by competing for the space. The vessels should also contain a minimal surface area in order to minimize possible adsorption sites.

Pretreatment of the sample is dependent upon the nature of the sample. See Table 4.1 for sample pretreatment methods.

Sample	Examples	Pretreatment method
Aqueous solutions	Water, beverages, urine, blood	Digestion if interference causing substituents are present
Suspensions	Water, beverages, urine, blood	Solid matter must either be re- moved by filtration, centrifugation or digestion, and then the methods for aqueous solutions can be followed
Organic liquids	Fuels, oils	Either direct measurement with AAS or dilution with organic material followed by measurement with AAS, standards must contain the analyte in the same form as the sample
Solids	Foodstuffs, rocks	Digestion followed by electrothermal AAS

Table 4.1: Sample pretreatment methods for AAS.

Calibration curve

In order to determine the concentration of the analyte in the solution, calibration curves can be employed. Using standards, a plot of concentration versus absorbance can be created. Three common methods used to make calibration curves are the standard calibration technique, the bracketing technique, and the analyte addition technique.

Standard calibration technique

This technique is the both the simplest and the most commonly used. The concentration of the sample is found by comparing its absorbance or integrated absorbance to a curve of the concentration of the standards versus the absorbances or integrated absorbances of the standards. In order for this method to be applied the following conditions must be met:

- Both the standards and the sample must have the same behavior when atomized. If they do not, the matrix of the standards should be altered to match that of the sample.
- The error in measuring the absorbance must be smaller than that of the preparation of the standards.
- The samples must be homogeneous.

The curve is typically linear and involves at least five points from five standards that are at equidistant concentrations from each other Figure 4.14.

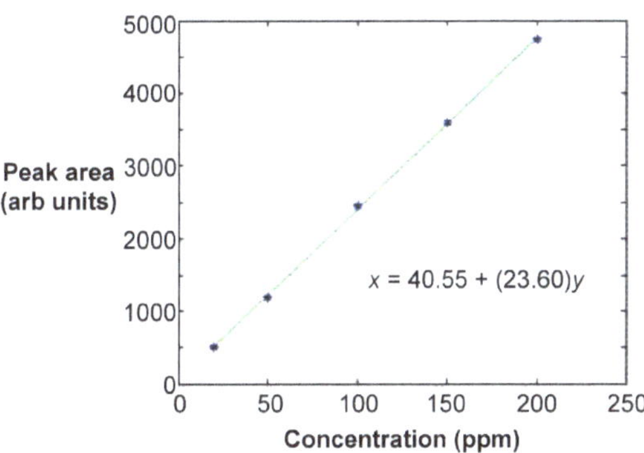

Figure 4.15: An example of a calibration curve made for the standard calibration technique.

This ensures that the fit is acceptable. A least means squares calculation is used to linearly fit the line. In most cases, the curve is linear only up to absorbance values of 0.5 to 0.8. The absorbance values of the standards should have the absorbance value of a blank subtracted.

Bracketing technique

The bracketing technique is a variation of the standard calibration technique. In this method, only two standards are necessary with concentrations c_1 and c_2. They bracket the approximate value of the sample concentration very closely. To determines the value for the sample, apply,

$$c_x = \frac{(A_x - A_1)(c_1 - c_2)}{A_2 - A_1} + c_1$$

where c_x and A_x are the concentration and absorbance of the unknown, and A_1 and A_2 are the absorbance for c_1 and c_2, respectively.

This method is very useful when the concentration of the analyte in the sample is outside of the linear portion of the calibration curve because the bracket is so small that the portion of the curve being used can be portrayed as linear. Although this method can be used accurately for nonlinear curves, the further the curve is from linear the greater the error will be. To help reduce this error, the standards should bracket the sample very closely.

Analyte addition technique

The analyte addition technique is often used when the concomitants in the sample are expected to create many interferences and the composition of the sample is unknown. The previous two techniques both require that the standards have a similar matrix to that of the sample, but that is not possible when the matrix is unknown. To compensate for this, the analyte addition technique uses an aliquot of the sample itself as the matrix. The aliquots are then spiked with various amounts of the analyte. This technique must be used only within the linear range of the absorbances.

Measurement interference

Interference is caused by contaminants within the sample that absorb at the same wavelength as the analyte, and thus can cause inaccurate measurements. Corrections can be made through a variety of methods such as background correction, addition of chemical additives, or addition of analyte Table 4.2.

Interference type	Cause of interference	Result	Correction measures
Atomic line overlap	Spectral profile of two elements are within 0.01 nm of each other	Higher experimental absorption value than the real value	Typically, doesn't occur in practical situations, so there is no established correction method
Molecular band and line overlap	Spectral profile of an element overlaps with molecular band	Higher experimental absorption value than the real value	Background correction
Ionization (vapor-phase or cation enhancement)	atoms are ionized at the temperature of the flame/ furnace, which decreases the amounts of free atoms	Lower experimental absorption value than real value	Add an ionization suppressor (or buffer) to both the sample and the standards
Light scattering	Solid particles scatter the beam of light lowering the intensity of the beam entering the monochromator	Higher experimental absorption value than the real value	Matrix modification and/or background correction
Chemical	The chemical being analyzed is contained within a compound in the analyte that is not atomized	Lower experimental absorption value than real value	Increase the temperature of the flame if flame AAS is being used, use a releasing chemical, or standard addition for electrothermal AAS
Physical	If physical properties of the sample and the standards are different, atomization can be affected thus affecting the number of free atom population	Can vary in either direction depending upon the conditions	Alter the standards to have similar physical properties to the samples
Volatilization	In electrothermal atomization, interference will occur if the rate of volatilization is not the same for the sample as for the standard, which is often caused by a heavy matrix	Can vary in either direction depending upon the conditions	Change the matrix by standard addition, or selectively volatilize components of the matrix

Table 4.2: Examples of interference in AAS.

Bibliography

L. Ebon, A. Fisher and S. J. Hill, *An Introduction to Analytical Atomic Spectrometry*, Ed. E. H. Evans, Wiley, New York (1998).

B. Welz and M. Sperling, *Atomic Absorption Spectrometry*, 3rd edn., Wiley-VCH, New York (1999).

J. W. Robinson, *Atomic Spectroscopy*, 2nd edn., Marcel Dekker, Inc., New York (1996).

K. S. Subramanian, V. S. Sastri, M. Elboujdaini, J. W. Connor, and A. B. C. Davey, Water contamination: Impact of tin-lead solder. *Water Res.*, 1995, **29**, 1827.

M. Sakata and O. Shimoda, A simple and rapid method for the determination of lead and cadmium in sediment by graphite furnace atomic absorption spectrometry. *Water Res.*, 1982, **16**, 231.

J. C. Van Loon, *Analytical Atomic Absorption Spectroscopy Selected Methods*, Academic Press, New York (1980).

Chapter 5: Inductively Coupled Atomic Emission Spectroscopy

Alvin Orbaek White and Andrew R. Barron

Introduction

Inductively coupled plasma atomic emission spectroscopy (ICP-AES) is a spectral method used to determine very precisely the elemental composition of samples; it can also be used to quantify the elemental concentration with the sample. ICP-AES uses high-energy plasma from an inert gas like argon to burn analytes very rapidly. The color that is emitted from the analyte is indicative of the elements present, and the intensity of the spectral signal is indicative of the concentration of the elements that is present. A schematic view of a typical experimental set-up is shown in Figure 5.1.

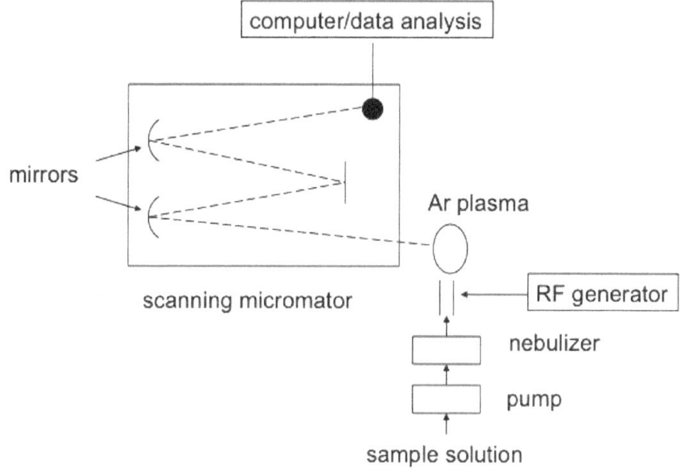

Figure 5.1: Schematic representation of an ICP-AES set-up.

How does ICP-AES work?

ICP-AES works by the emission of photons from analytes that are brought to an excited state by the use of high-energy plasma. The plasma source is induced when passing argon gas through an alternating electric field that is created by an inductively couple coil. When the analyte is excited the electrons try to dissipate the induced energy moving to a ground state of lower energy, in doing this they emit the excess energy in the form of light. The wavelength of light emitted depends on the energy gap between the excited energy level and the ground state. This is specific to the element based on the number of electrons the element has and electron orbitals are filled. In this way the wavelength of light can be used to determine what elements are present by detection of the light at specific wavelengths.

As a simple example consider the situation when placing a piece of copper wire into the flame of a candle. The flame turns green due to the emission of excited electrons within the copper metal, as the electrons try to dissipate the energy incurred from the flame, they move to a more stable state emitting energy in the form of light. The energy gap between the excited state to the ground state (ΔE) dictates the color of the light or wavelength of the light,

$$E = h\nu$$

where h is Plank's constant (6.626×10^{-34} m^{2}kg/s), and ν is the frequency of the emitted light.

The wavelength of light is indicative of the element present. If another metal is placed in the flame such as iron a different color flame will be emitted because the electronic structure of iron is different from that of copper. This is a very simple analogy for what is happening in ICP-AES and how it is used to determine what elements are present. By detecting the wavelength of light that is emitted from the analyte one can deduce what elements are be present.

Naturally if there is a lot of the material present then there will be an accumulative effect making the intensity of the signal large. However, if there were very little materials present the signal would be low. By this rationale one can create a calibration curve from analyte solutions of known concentrations, whereby the intensity of the signal changes as a function of the concentration of the material that is present. When measuring the intensity from a sample of unknown concentration the intensity from this sample can be compared to that from the calibration curve, so this can be used to determine the concentration of the analytes within the sample.

Determine of nanoparticle elemental composition

As with any sample being studied by ICP-AES nanoparticles need to be digested so that all the atoms can be vaporized in the plasma equally. If a metal containing nanoparticle were not digested using a strong acid to bring the metals atoms into solution, the form of the particle could hinder some of the material being vaporized. The analyte would not be detected even though it is present in the sample and this would give an erroneous result. Nanoparticles are often covered with a protective layer of organic ligands and this must be removed also. Further to this the solvent used for the nanoparticles may also be an organic solution and this should be removed as it too will not be miscible in the aqueous medium.

Several organic solvents have low vapor pressures, so it is relatively easy to remove the solvent by heating the samples, removing the solvent by evaporation. To remove the organic ligands that are present on the nanoparticle, choric acid can be used. This is a very strong acid and can break down the organic ligands readily. To digest the particles and get the metal into solution concentrated nitric acid (HNO_3) is most often used.

A typical protocol may use 0.5 mL of concentrated nanoparticle solution and digest this with 9.5 mL of concentrated nitric acid over the period of a few days. After which 0.5 mL of the digested solution is placed in 9.5 mL of ultrapure water with a resistance of greater than 18.2 MΩ/cm is desirable. Nanopure or MilliQ water are named after the systems that produce them. The reason why ultrapure water is used is because DI water or regular water will have some amount of metals ions present and these will be detected by the ICP-AES measurement and will lead to figures that are not truly representative of the analyte concentration alone. This is especially pertinent when there is a very a low concentration of metal analyte to be detected and is even more a problem when the metal to be detected is commonly found in water such as iron. Once the ultrapure water and digested solution are prepared then the sample is ready for analysis.

Another point to consider when doing ICP-AES on nanoparticles to determine chemical compositions, includes the potential for wavelength overlap. The energy that is released in the form of light is unique to each element, but elements that are very similar in atomic structure will have emission wavelengths that are very similar to one another. Consider the example of iron and cobalt. Iron has an emission wavelength at 238.204 nm and cobalt has an emission wavelength at 238.892 nm. So, if you were to try to determine the amount of each element in an alloy of the two you would have to select another wavelength that would be unique to that element, and not have any wavelength overlap to other analytes in solution. For this case of iron and cobalt it would be wiser to use a wavelength for iron detection of 259.940 nm and a wavelength detection of 228.616 nm. Bearing this in mind a good rule of thumb is to try use the wavelength of the analyte that affords the best detection primarily. But if this value leads to a possible wavelength overlap of within 15 nm wavelength with another analyte in the solution then another choice should be made of the detection wavelength to prevent wavelength overlap from occurring.

Some people have also used the ICP-AES technique to determine the size of nanoparticles. The signal that is detected is determined by the amount of the material that is present in solution. If very dilute solutions of nanoparticles are being analyzed, particles are being analyzed one at a time, i.e., there will be one nanoparticle per droplet in the nebulizer. The signal intensity would then differ according to the size of the particle. In this way the ICP-AES technique could be used to determine the concentration of the particles in the solution as well as the size of the particles.

Calculations for ICP concentrations

In order to perform ICP-AES stock solutions must be prepared in dilute nitric acid solutions. To do this a concentrated solution should be diluted with D water to prepare 7 wt% nitric acid solutions. If the concentrated solution is 69.8 wt% (check the assay amount that is written on the side of the bottle) then the amount to dilute the solution will be as such:
- the density (d) of HNO_3 is 1.42 g/mL,
- molecular weight (M_w) of HNO_3 is 63.01.

Concentrated percentage 69.8 wt% from assay. First you must determine the molarity of the concentrated solution,

$$\text{Molarity} = [(\%)(d)/(M_w)] * 10$$

For the present assay amount, the figure will be calculated from:

$$M = [(69.8)(1.42) / (63.01)] * 10$$

$$\therefore M = 15.73$$

We can define the unknowns as follows:

- C_I = concentration of concentrated solution (ppm),
- C_F = desired concentration (ppm),
- M_I = initial mass of material (mL),
- M_F = mass of material required for dilution (mL).

This is the initial concentration C_I. To determine the molarity of the 7% solution will be determined by

$$M = [(7)(1.42) / (63.01)] * 10$$

i.e., is the final concentration C_F.

$$\therefore M = 1.58$$

We use these figures in to determine the amount of dilution required to dilute the concentrated nitric acid to make it a 7% solution.

$$mass_I * concentration_I = mass_F * concentration_F$$

Now as we are talking about solutions the amount of mass will be measured in mL, and the concentration will be measured as a molarity,

$$mL_I * C_I = mL_F * C_F$$

$$\therefore mL_I = [mL_F * C_F]/C_I$$

where M_I and M_F have been calculated above. In addition, the amount of dilute solution will be dependent on the user and how much is required by the user to complete the ICP analysis, for the sake of argument let's say that we need 10 mL of dilute solution, this is mL_F,

$$mL_I = [10 * 1.58] / 15.73$$

$$\therefore mL_I = 10.03 \ mL$$

This means that 10.03 mL of the concentrated nitric acid (69.8%) should be diluted up to a total of 100 mL with DI water.

Now that you have your stock solution with the correct percentage then you can use this solution to prepare your solutions of varying concentration. Let's take the example that the stock solution that you purchase from a supplier has a concentration of 100 ppm of analyte, which is equivalent to 1 μg/mL.

In order to make your calibration curve more accurate it is important to be aware of two issues. Firstly, as with all straight-line graphs, the more points that are used then the better the statistics is that the line is correct. But, secondly, the more measurements that are used means that more room for error is introduced to the system, to avoid these errors from occurring one should be very vigilant and skilled in the use of pipetting and diluting of solutions. Especially when working with very low concentration solutions a small drop of material making the dilution above or below the exactly required amount can alter the concentration and hence affect the calibration deleteriously.

The choice of concentrations to make will depend on the samples and the concentration of analyte within the samples that are being analyzed. For first time users it is wise to make a calibration curve with a large range to encompass all the possible outcomes. When the user is more aware of the kind of concentrations that they are producing in their synthesis then they can narrow down the range to t the kind of concentrations that they are anticipating.

In this example we will make concentrations ranging from 10 ppm to 0.1 ppm, with a total of five samples. In a typical ICP-AES analysis about 3 mL of solution is used, however if you have situations with substantial wavelength overlap then you may have chosen to do two separate runs and so you will need approximately 6 mL solution. In general, it is wise to have at least 10 mL of solution to prepare for any eventuality that may occur. There will also be some extra amount needed for samples that are being used for the quality control check. For this reason, 10 mL should be a sufficient amount to prepare of each concentration.

The methodology adopted works as follows. Make the high concentration solution then take from that solution and dilute further to the desired concentrations that are required.

Let's say the concentration of the stock solution from the supplier is 100 ppm of analyte. First, we should dilute to a concentration of 10 ppm. To make 10 mL of 10 ppm solution we should take 1 mL of the 100 ppm solution and dilute it up to a volume of 10 mL with ultrapure water, now the concentration of this solution is 10 ppm. Then we can take from the 10 ppm solution and dilute this down to get a solution with 5 ppm. To do this take 5 mL of the 10 ppm solution and dilute it to 10 mL with ultrapure water, then you will have a solution of 10 mL that is 5 ppm concentration. And so, you can do this successively taking aliquots from each solution working your way down at incremental steps until you have a series of solutions that have concentrations ranging from 10 ppm all the way down to 0.1 ppm or lower, as required.

ICP-AES: a case study

While ICP-AES is a useful method for quantifying the presence of a single metal in a given nanoparticle, another very important application comes from the ability to determine the ratio of metals within a sample of nanoparticles.

In the following examples we can consider the bi-metallic nanoparticles of iron with copper. In a typical synthesis 0.75 mmol of $Fe(acac)_3$ is used to prepare iron-oxide nanoparticle of the form Fe_3O_4. It is possible to replace a quantity of the Fe^{n+} ions with another metal of similar charge. In this manner bi-metallic particles were made with a precursor containing a suitable metal. In this example the additional metal precursor will be $Cu(acac)_2$.

Keep the total metal concentration in this example is 0.75 mmol. So, if we want to see the effect of having 10% of the metal in the reaction as copper, then we will use 10% of 0.75 mmol, that is 0.075 mmol $Cu(acac)_2$, and the corresponding amount of iron is 0.675 mmol $Fe(acac)_3$. We can do this for successive increments of the metals until you make 100% copper oxide particles.

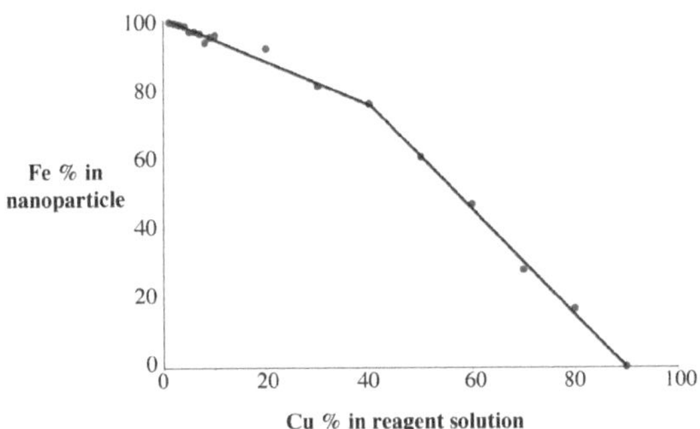

Figure 5.2: Change in iron percentage in the Fe-Cu-O nanoparticles as a function of how much iron precursor is used in the synthesis of the nanoparticles.

Subsequent Fe and Cu ICP-AES of the samples will allow the determination of Fe:Cu ratio that is present in the nanoparticle. This can be compared to the ratio of Fe and Cu that was applied as reactants. The graph Figure 5.2 shows how the percentage of Fe in the nanoparticle changes as a function of how much Fe is used as a reagent.

Determining analyte concentration

Once the nanoparticles are digested and the ICP-AES analysis has been completed you must turn the figures from the ICP-AES analysis into working numbers to determine the concentration of metals in the solution that was synthesized initially.

Let's first consider the nanoparticles that are of one metal alone. The figure given by the analysis in this case is given in units of mg/L, this is the value in ppm's. This figure was recorded for the solution that was analyzed, and this is of a dilute concentration compared to the initial synthesized solution because the particles had to be digested in acid first, then diluted further into DI water.

As mentioned above in the experimental 0.5 mL of the synthesized nanoparticles were first digested in 9.5 mL of concentrated nitric acid. Then when the digestion was complete 0.5 mL of this solution was dissolved in 9.5 mL of DI water. This was the final solution that was analyzed using ICP, and the concentration of metal in this solution will be far lower than that of the original solution. In this case the amount of analyte in the final solution being analyzed is $1/20^{th}$ that of the total amount of material in the solution that was originally synthesized.

Calculating concentration in ppm

Let us take an example that upon analysis by ICP-AES the amount of Fe detected is 6.38 mg/L. First convert the figure to mg/mL, using

$$6.38 \text{ mg/L} * 1/1000 \text{ L/mL} = 6.38\text{x}10^{-3} \text{ mg/mL}$$

The amount of material was diluted to a total volume of 10 mL. Therefore, we should multiply this value by 10 mL to see how much mass was in the whole container,

6.38×10^{-3} mg/mL * 10 mL = 6.38×10^{-2} mg

This is the total mass of iron that was present in the solution that was analyzed using the ICP device. To convert this amount to ppm we should take into consideration the fact that 0.5 mL was initially diluted to 10 mL, to do this we should divide the total mass of iron by this amount that it was diluted to,

6.38×10^{-2} mg / 0.5 mL = 0.1276 mg/mL

This was the total amount of analyte in the 10 mL solution that was analyzed by the ICP device, to attain the value in ppm it should be multiplied by a thousand, that is then 127.6 ppm of Fe.

Determining concentration of original solution

We now need to factor in the fact that there were several dilutions of the original solution first to digest the metals and then to dissolve them in DI water, in all there were two dilutions and each dilution was equivalent in mass. By diluting 0.5 mL to 10 mL, we are effectively diluting the solution by a factor of 20, and this was carried out twice,

0.1276 mg/mL * 20 = 2.552 mg/mL

This is the amount of analyte in the solution of digested particles, to covert this to ppm we should multiply it by 1/1000 mL/L, in the following way:

2.552 mg/mL * 1/1000 mL/L = 2552 mg/L

This is essentially your answer now as 2552 ppm. This is the amount of Fe in the solution of digested particles. This was made by diluting 0.5 mL of the original solution into 9.5 mL concentrated nitric acid, which is the same as diluting by a factor of 20. To calculate how much analyte was in the original batch that was synthesized we multiply the previous value by 20 again,

2552 ppm * 20 = 51040 ppm

This is the final amount of Fe concentration of the original batch when it was synthesized and made soluble in hexanes.

Calculating stoichiometric ratio

Moving from calculating the concentration of individual elements now we can concentrate on the calculation of stoichiometric ratios in the bi-metallic nanoparticles.

Consider the case when we have the iron and the copper elements in the nanoparticle. The amounts determined by ICP are:

- Iron = 1.429 mg/L,
- Copper = 1.837 mg/L.

We must account for the molecular weights of each element by dividing the ICP obtained value, by the molecular weight for that particular element. For iron this is calculated by,

$$1.429 \text{ mg/L} / 55.85 = 0.0211$$

and thus, this is molar ratio of iron. On the other hand, the ICP returns a value for copper that is given by

$$1.837 \text{ mg/L} / 63.55 = 0.0289$$

To determine the percentage of iron we use

$$\%Fe = [(\text{molar ratio of iron})/(\text{sum of molar ratios})] * 100$$

which gives a percentage value of 42.15% Fe.

To work out the copper percentage we calculate this amount using (1.65), which leads to an answer of 57.85% Cu.

$$\% Cu = [(\text{molar ratio of copper})/(\text{sum of molar ratios})] * 100$$

In this way the percentage iron in the nanoparticle can be determined as function of the reagent concentration prior to the synthesis (Figure 5.2).

Determining concentration of nanoparticles in solution

The previous examples have shown how to calculate both the concentration of one analyte and the effective shared concentration of metals in the solution. These figures pertain to the concentration of elemental atoms present in solution. To use this to determine the concentration of nanoparticles we must first consider how many atoms that are being detected are in a nanoparticle. Let us consider that the Fe_3O_4 nanoparticles are of 7 nm diameter. In a 7 nm particle we expect to find 20,000 atoms. In this analysis, however, we have only detected Fe atoms, so we must still account for the number of oxygen atoms that form the crystal lattice also.

For every 3 Fe atoms, there are 4 O atoms. But as iron is slightly larger than oxygen, it will make up for the fact there is one less Fe atom. This is an over-simplification but at this time it serves the purpose to make the reader aware of the steps that are required to take when judging nanoparticles concentration. Let us consider that half of the nanoparticle size is attributed to iron atoms, and the other half of the size is attributed to oxygen atoms.

As there are 20,000 atoms total in a 7 nm particle, and then when considering the effect of the oxide state we will say that for every 10,000 atoms of Fe you will have a 7 nm particle. So now we must find out how many Fe atoms are present in the sample so we can divide by 10,000 to determine how many nanoparticles are present.

In the case from above, we found the solution when synthesized had a concentration 51,040 ppm Fe atoms in solution. To determine how many atoms this equates to we will use the fact that 1 mole of material has the Avogadro number of atoms present,

51040 ppm = 51040 mg/L = 51.040 g/L

1 mole of iron weighs 55.847 g. To determine how many moles we now have, we divide the values like this:

(51.040 g/L) / (55.847 g) = 0.9139 moles/L

The number of atoms is found by multiplying this by Avogadro's number (6.022×10^{23}):

(0.9139 moles/L) * (6.022×10^{23} atoms) = 5.5×10^{23} atoms/L

For every 10,000 atoms we have a nanoparticle (NP) of 7 nm diameter, assuming all the particles are equivalent in size we can then divide the values,

(5.5×10^{23} atoms/L) / (10,000 atoms/NP) = 5.5×10^{19} NP/L

This is the concentration of nanoparticles per liter of solution as synthesized.

Combined surface area

One very interesting thing about nanotechnology that nanoparticles can be used for is their incredible ratio between the surface areas compared with the volume. As the particles get smaller and smaller the surface area becomes more prominent. And as much of the chemistry is done on surfaces, nanoparticles are good contenders for future use where high aspect ratios are required.

In the example above we considered the particles to be of 7 nm diameters. The surface area of such a particle is 1.539×10^{-16} m^2. So, the combined surface area of all the particles is found by multiplying each particle by their individual surface areas.

(1.539×10^{-16} m^2) * (5.5×10^{19} NP/L) = 8465 m^2/L

To put this into context, an American football field is approximately 5321 m^2. So, a liter of this nanoparticle solution would have the same surface area of approximately 1.5 football fields. That is a lot of area in one liter of solution when you consider how much material it would take to line the football field with thin layer of metallic iron. Remember there is only about 51 g/L of iron in this solution!

Bibliography

C. A. Crouse and A. R. Barron, Reagent control over the size, uniformity, and composition of Co–Fe–O nanoparticles. *J. Mater. Chem.*, 2008, **18**, 4146.

S. de Villiers, M. Greaves, and H. Elderfield, An intensity ratio calibration method for the accurate determination of Mg/Ca and Sr/Ca of marine carbonates by ICP-AES. *Geochem Geophys.*, 2003, **4**, 8406.

C. Moor, T. Lymberopoulou, and V. J. Dietrich, Determination of heavy metals in soils, sediments and geological materials by ICP-AES and ICP-MS. *Microchimica Acta*, 2001, **136**, 123.

S. Sun and H. Zeng, Size-controlled synthesis of magnetite nanoparticles. *J. Am. Chem. Soc.*, 2002, **124**, 8204.

J. L. Todolí, L. Gras, V. Hernandis, and J. Mora, Elemental matrix effects in ICP-AES. *J. Anal. At. Spectrom.*, 2002, **17**, 142.

Chapter 6: ICP-MS for Trace Metal Analysis

Meghan Jebb and Andrew R. Barron

Introduction

Inductively coupled plasma mass spectroscopy (ICP-MS) is an analytical technique for determining trace multi-elemental and isotopic concentrations in liquid, solid, or gaseous samples. It combines an ion generating argon plasma source with the sensitive detection limit of mass spectrometry detection. Although ICP-MS is used for many different types of elemental analysis, including pharmaceutical testing and reagent manufacturing, this module will focus on its applications in mineral and water studies. Although akin to ICP-AES (inductively coupled plasma atomic emission spectroscopy), ICP-MS has significant differences.

Basic instrumentation and operation

As shown in Figure 6.1 there are several basic components of an ICP-MS instrument, which consist of a sampling interface, a peristaltic pump leading to a nebulizer, a spray chamber, a plasma torch, a detector, an interface, and ion-focusing system, a mass-separation device, and a vacuum chamber, maintained by turbo molecular pumps.

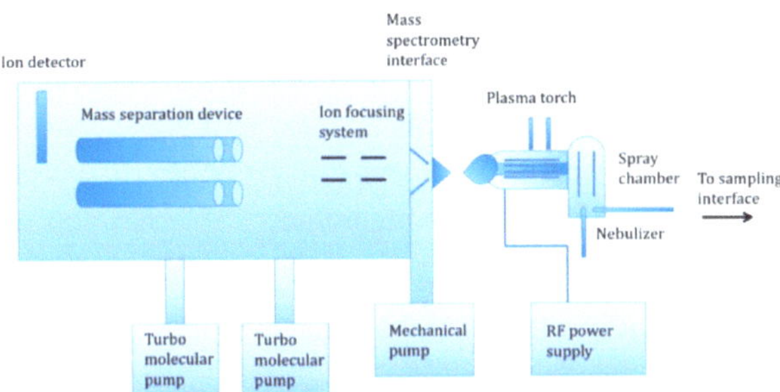

Figure 6.1: Scheme depicting the basic components of an ICP-MS system. Adapted from R. Thomas, *Practical Guide to ICP-MS: A Tutorial for Beginners*, 2nd edn., CRC Press, Boca Raton (2008).

The basic operation works as follows: a liquid sample is pumped into the nebulizer to convert the sample into a spray. An internal standard, such as germanium, is pumped into a mixer along with the sample prior to nebulization to compensate for matrix effects. Large droplets are filtered out, and small droplets continue into the plasma torch, turning to ions. The mass separation device separates these ions based on their mass-to-charge ratio. An ion detector then converts these ions into an electrical signal, which is multiplied and read by computer software.

The main difference between ICP-MS and ICP-AES (Table 6.1) is the way in which the ions are generated and detected. In ICP-AES, the ions are excited by vertical plasma, emitting photons that are separated on the basis of their emission wavelengths. As implied by the name, ICP-MS separates the ions, generated by horizontal plasma, on the basis of their mass-to-charge ratios (m/z). In fact, caution is taken to prevent photons from reaching the detector and creating background noise. The difference in ion formation and detection methods has a significant impact on the relative sensitivities of the two techniques. While both methods are capable of very fast, high throughput multi-elemental analysis (10 - 40 elements per minute per sample), ICP-MS has a detection limit of a few ppt to a few hundred ppm, compared to the ppb-ppm range (1 ppb - 100 ppm) of ICP-AES. ICP-MS also works over eight orders of magnitude detection level compared to ICP-AES' six. As a result of its lower sensitivity, ICP-MS is a more expensive system. One other important difference is that only ICP-MS can distinguish between different isotopes of an element, as it segregates ions based on mass.

	ICP-MS	ICP-AES
Plasma	Horizontal: generates cations	Vertical: excites atoms, which emit photons
Ion detection	Mass-to-charge ratio	Wavelength of emitted light
Detection limit	1-10 ppt	1-10 ppb
Working range	8 orders of magnitude	6 orders of magnitude
Throughput	20-30 elements per minute	10-40 elements per minute
Isotope detection	Yes	No
Cost	~$150,000	~$50,000
Multi-element detection	Yes	Yes
Spectral interferences	Predictable, less than 300	Much greater in number and more complicated to correct
Routine accessories	Electrothermal vaporization, laser ablation, high-performance liquid chromatography, etc.	Rare

Table 6.1: Comparison of ICP-MS and ICP-AES.

Sample preparation

With such small sample sizes, care must be taken to ensure that collected samples are representative of the bulk material. This is especially relevant in rocks and minerals, which can vary widely in elemental content from region to region. Random, composite, and integrated sampling are each different approach for obtaining representative samples.

Because ICP-MS can detect elements in concentrations as minute as a few nanograms per liter (parts per trillion), contamination is a very serious issue associated with collecting and storing samples prior to measurements. In general, use of glassware should be minimized, due to leaching impurities from the glass or absorption of analyte by the glass. If glass is used, it should be washed periodically with a strong oxidizing agent, such as dichromic acid ($H_2Cr_2O_7$, Figure 6.2), or a commercial glass detergent. In terms of sample containers, plastic is usually better than glass, polytetrafluoroethylene (PTFE, Teflon®, Figure 6.3) being regarded as the cleanest plastics; however, even these materials can contain leachable contaminants, such as phosphorus or barium compounds. All containers, pipettes, pipette tips, and the like should be soaked in 1 - 2% HNO_3. Nitric acid is preferred over HCl, which can ionize in the plasma to form $^{35}Cl^{16}O^+$ and $^{40}Ar^{35}Cl^{3+}$, which have the same mass-to-charge ratios as $^{51}V^+$ and $^{75}As^+$, respectively. If possible, samples should be prepared as close as possible to the ICP-MS instrument without being in the same room.

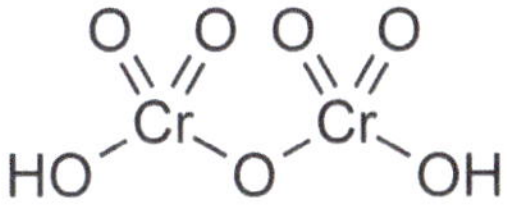

Figure 6.2: The structure of dichromic acid

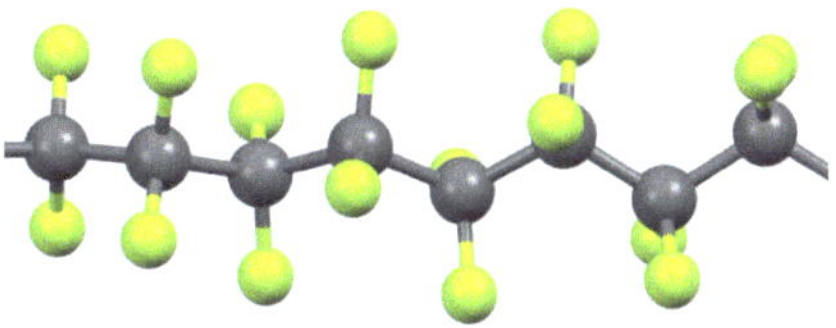

Figure 6.3: A ball and stick model of polytetrafluoroethylene (PTFE): C = grey and F = green.

With the exception of solid samples analyzed by laser ablation ICP-MS, samples must be in liquid or solution form. Solids are ground into a fine powder with a mortar and pestle and passed through a mesh sieve. Often the first sample is discarded to prevent contamination from the mortar or sieve. Powders are then digested with ultrapure concentrated acids or oxidizing agents, like chloric acid ($HClO_3$), and diluted to the correct order of magnitude with 1 - 2% trace metal grade nitric acid.

Once in liquid or solution form, the samples must be diluted with 1 - 2% ultrapure HNO_3 to a low concentration to produce a signal intensity lower than about 10^6 counts. Not all elements have the same concentration to intensity correlation; therefore, it is safer to test unfamiliar samples on ICP-AES first. Once properly diluted, the sample should be filtered through a 0.25 - 0.45 μm membrane to remove particulates.

Gaseous samples can also be analyzed by direct injection into the instrument. Alternatively, gas chromatography equipment can be coupled to an ICP-MS machine for separation of multiple gases prior to sample introduction.

Standards

Multi- and single-element standards can be purchased commercially and must be diluted further with 1 - 2% nitric acid to prepare different concentrations for the instrument to create a calibration curve, which will be read by the computer software to determine the unknown concentration of the sample. There should be several standards, encompassing the expected concentration of the sample. Completely unknown samples should be tested on less sensitive instruments, such as ICP-AES or EDXRF (energy dispersive X-ray fluorescence), before ICP-MS.

Limitations of ICP-MS

While ICP-MS is a powerful technique, users should be aware of its limitations. Firstly, the intensity of the signal varies with each isotope, and there is a large group of elements that cannot be detected by ICP-MS. This consists of H, He and most gaseous elements, C, and elements without naturally occurring isotopes, including most actinides.

There are many different kinds of interferences that can occur with ICP-MS, when plasma-formed species have the same mass as the ionized analyte species. These interferences are predictable and can be corrected with element correction equations or by evaluating isotopes with lower natural abundances. Using a mixed gas with the argon source can also alleviate the interference.

The accuracy of ICP-MS is highly dependent on the user's skill and technique. Standard and sample preparations require utmost care to prevent incorrect calibration curves and contamination. As exempli ed below, a thorough understanding of chemistry is necessary to predict conflicting species that can be formed in the plasma and produce false positives. While an inexperienced user may be able to obtain results fairly easily, those results may not be trustworthy. Spectral interference and matrix effects are problems that the user must work diligently to correct.

Applications: analysis of mineral and water samples

In order to illustrate the capabilities of ICP-MS, various geochemical applications as described. The chosen examples are representative of the types of studies that rely heavily on ICP-MS, highlighting its unique capabilities.

Trace elemental analysis of minerals

With its high throughput, ICP-MS has made sensitive analysis of multi-element detection in rock and mineral samples feasible. Studies of trace components in rock can reveal information about the chemical evolution of the mantle and crust. For example, spinel peridotite xenoliths (Figure 6.4), which are igneous rock fragments derived from the mantle, were analyzed for 27 elements, including lithium, scandium and titanium at the parts per million level and yttrium, lutetium, tantalum, and hafnium in parts per billion. X-ray fluorescence was used to complement ICP-MS, detecting metals in bulk concentrations. Both liquid and solid samples were analyzed, the latter being performed using laser-ablation ICP-MS, which points out the suitability of the technique for being used in tandem with others. In order to prepare the solution samples, optically pure minerals were sonicated in 3 M HCl, then 5% HF, then 3 M HCl again and dissolved in distilled water. The solid samples were converted into plasma by laser ablation prior to injection into the nebulizer of the LA-ICP-MS instrument. The results showed good agreement between the laser ablation and solution methods. Furthermore, this comprehensive study shed light on the partitioning behavior of incompatible elements, which, due to their size and charge, have difficulty entering cation sites in minerals. In the upper mantle, incompatible trace elements, especially barium, niobium and tantalum, were found to reside in glass pockets within the peridotite samples.

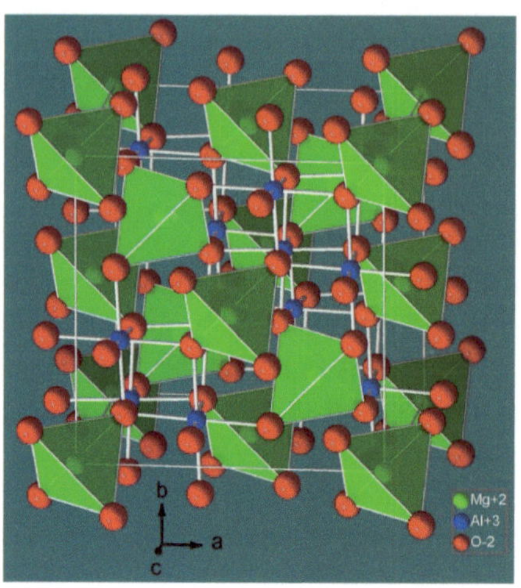

Figure 6.4: Crystal structure of a typical spinel, general formula $A^{2+}B_2^{3+}O_4^{3-}$.

Trace elemental analysis of water

Another important area of geology that requires knowledge of trace elemental compositions is water analysis. In order to demonstrate the full capability of ICP-MS as an analytical technique in this field, researchers aim to use the identification of trace metals present in groundwater to determine a fingerprint for a particular water source. In one study the analysis of four different Nevada springs determined trace metal analysis in parts per billion and even parts per trillion (ng/L). Because they were present is such low concentrations, samples containing rare earth elements lutetium, thulium, and terbium were preconcentrated by a cation exchange column to enable detection at 0.05 ppt.

For some isotopes, special corrections necessary to account for false positives, which are produced by plasma-formed molecules with the same mass-to-charge ratio as the isotopic ions. For instance, false positives for Sc ($m/z = 45$) or Ti ($m/z = 47$) could result from CO_2H^+ ($m/z = 45$) or PO^+ ($m/z = 47$); and BaO^+ ($m/z = 151, 153$) conflicts with Eu-151 and Eu-153. In the latter case, barium has many isotopes (134, 135, 136, 137, 138) in various abundances, Ba-138 comprising 71.7% barium abundance. ICP-MS detects peaks corresponding to BaO^+ for all isotopes. Thus, researchers were able to approximate a more accurate europium concentration by monitoring a non-interfering barium peak and extrapolating back to the concentration of barium in the system. This concentration was subtracted out to give a more realistic europium concentration. By employing such strategies, false positives could be taken into account and corrected. Additionally, 10 ppb internal standard was added to all samples to correct for changes in sample matrix, viscosity and salt buildup throughout collection. In total, 54 elements were detected at levels spanning seven orders of magnitude. This study demonstrates the incredible sensitivity and working range of ICP-MS.

Determination of arsenic content

Elemental analysis in water is also important for the health of aquatic species, which can ultimately affect the entire food chain, including people. With this in mind, arsenic content was determined in fresh water and aquatic organisms in Hayakawa River in Kanagawa, Japan (Figure 6.5), which has very high arsenic concentrations due to its hot spring source in Owakudani Valley. While water samples were simply filtered and prior to analysis, organisms required special preparation, in order to be compatible with the sampler. Organisms collected for this studied included water bug, green macroalga, fish, and crustaceans.

Figure 6.5: Oharabashi Bridge over the Hayakawa River in Kanagawa, Japan.

For total arsenic content determination, the samples were freeze-dried to remove all water from the sample in order to know the exact final volume upon resuspension. Next, the samples were ground into a powder, followed by soaking in nitric acid, heating at 110 °C. The sample then underwent heating with hydrogen peroxide, dilution, and filtering through a 0.45 μm membrane.

This protocol served to oxidize the entire sample and remove large particles prior to introduction into the ICP-MS instrument. Samples that are not properly digested can build up on the plasma torch and cause expensive damage to the instrument. Since the plasma converts the sample into various ion constituents, it is unnecessary to know the exact oxidized products prior to sample introduction. In addition to total arsenic content, the arsenic concentration of different organic arsenic-containing compounds (arsenicals) produced in the organisms was measured by high performance liquid chromatography coupled to ICP-MS (HPLC/ICP-MS). The arsenicals were separated by HPLC before travelling into the ICP-MS instrument for arsenic concentration determination. For this experiment, the organic compounds were extracted from biological samples by dissolving freeze-dried samples in methanol/water solutions, sonicating, and centrifuging. The extracts were dried under vacuum, re-dissolved in water, and filtered prior to loading. This did not account for all compounds, however, because over 50% arsenicals were non-soluble in aqueous solution. One important plasma side product to account for was $ArCl^+$, which has the same mass-to-charge ratio ($m/z = 75$) as As. This was corrected by oxidizing the arsenic ions within the mass separation device in the ICP-MS vacuum chamber to generate AsO^+, with $m/z = 91$. The total arsenic concentration of the samples ranged from 17 - 18 ppm.

Bibliography

R. Thomas, *Practical Guide to ICP-MS: A Tutorial for Beginners*, 2nd edn., CRC Press, Boca Raton (2008).

A. Scheffer, C. Engelhard, M. Sperling, and W. Buscher, ICP-MS as a new tool for the determination of gold nanoparticles in bioanalytical applications. *Anal. Bioanal. Chem.*, 2008, **390**, 249.

K. J. Stetzenbach, M. Amano, D. K. Kreamer, and V. F. Hodge, Testing the limits of ICP-MS: determination of trace elements in ground water at the part-per-trillion level. *Groundwater*, 1994, **32**, 976.

S. M. Eggins, R. L. Rudnick, and W. F. McDonough, The composition of peridotites and their minerals: a laser-ablation ICP–MS study. *Earth Planet. Sci. Lett.*, 1998, **154**, 53.

S. Miyashita, M. Shimoya, Y. Kamidate, T. Kuroiwa, O. Shikino, S. Fujiwara, K. A. Francesconi, and T. Kaise, Rapid determination of arsenic species in freshwater organisms from the arsenic-rich Hayakawa River in Japan using HPLC-ICP-MS. *Chemosphere*, 2009, **75**, 1065.

Chapter 7: Ion Selective Electrode Analysis

Huilong Fei and Andrew R. Barron

Introduction

Ion selective electrode (ISE) is an analytical technique used to determine the activity of ions in aqueous solution by measuring the electrical potential. ISE has many advantages compared to other techniques, including:

- It is relatively inexpensive and easy to operate.
- It has wide concentration measurement range.
- As it measures the activity, instead of concentration, it is particularly useful in biological/medical application.
- It is a real-time measurement, which means it can monitor the change of activity of ion with time.
- It can determine both positively and negatively charged ions.

Based on these advantages, ISE has wide variety of applications, which is reasonable considering the importance of measuring ion activity. For example, ISE finds use in pollution monitoring of natural waters (CN^-, F^-, S^-, Cl^-, etc.), food processing (NO_3^-, NO_2^- in meat preservatives), Ca^{2+} analysis in dairy products, and K^+ analysis in fruit juices, etc.

Measurement setup

Before focusing on how ISE works, it would be good to get an idea what ISE setup looks like and the component of the ISE instrument. Figure 7.1 shows the basic components of ISE setup. It has an ion selective electrode, which allows measured ions to pass but excludes the passage of the other ions. Within this ion selective electrode, there is an internal reference electrode (Figure 7.1), which is made of silver wire coated with solid silver chloride, embedded in concentrated potassium chloride solution (filling solution) saturated with silver chloride. This solution also contains the same ions as that to be measured. There is also a reference electrode similar to ion selective electrode, but there is no to-be-measured ion in the internal electrolyte and the selective membrane is replaced by porous frit, which allows the slow passage of the internal filling solution and forms the liquid junction with the external text solution. The ion selective electrode and reference electrode are connected by a milli-voltmeter. Measurement is accomplished simply by immersing the two electrodes in the same test solution.

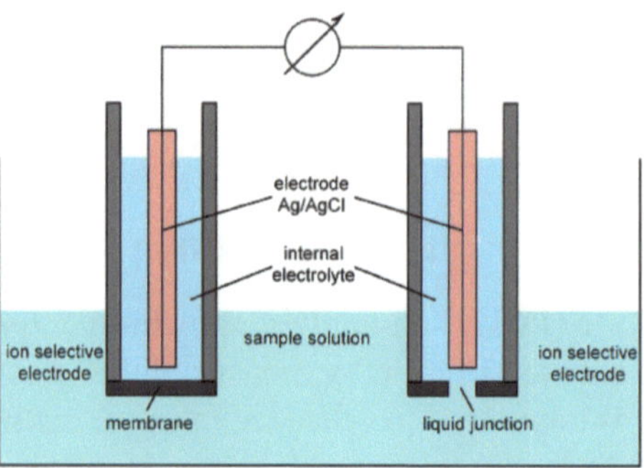

Figure 7.1: Measurement setup of ISE.

The theory of ISE

There are commonly more than one types of ions in solution. So how ISE manage to measure the concentration of certain ion in solution without being affected by other ions? This is done by applying a selective membrane at the ion selective electrode, which only allows the desired ion to pass through. At equilibrium, there is potential difference existing between two sides of the membrane, and it is governed by the concentration of the tested solution described by Nernst equation,

$$E = E_0 + (2.030 \ RT/nF)\log C$$

where E is potential, E_0 is a constant characteristic of a particular ISE, R is the gas constant (8.314 J/K.mol), T is the temperature (in K), n is the charge of the ion and F is Faraday constant (96,500 coulombs/mol).

To make it relevant, the measured potential difference is proportional to the logarithm of ion concentration. Thus, the relationship between potential difference and ion concentration can be determined by measuring the potential of two solutions of already-known ion concentration and a plot based on the measured potential and logarithm of the ion concentration. Based on this plot, the ion concentration of an unknown solution can be known by measuring the potential and corresponding it to the plot.

Determination of fluoride ion

Fluoride is added into drinking water and toothpaste to prevent dental caries and thus the determination of its concentration is of great importance to human health. Here, we will give some data and calculations to show how the concentration of fluoride ion is determined and have a glance at how relevant ISE is to our daily life. According to Nernst equation, in this case n = 1, T = 25 °C and E_0, R, F are constants and thus this equation can be simplified,

$$E = K + S \log C$$

The first step is to obtain a calibration curve for fluoride ion, and this can be done by preparing several fluoride standard solutions with known concentration (Table 7.1) and making a plot of E versus log C (Figure 7.2).

Concentration (mg/L)	log C	E (mV)
200.0	2.301	-35.6
100.0	2.000	-17.8
50.00	1.699	0.4
25.00	1.398	16.8
12.50	1.097	34.9
6.250	0.796	52.8
3.125	0.495	70.4
1.563	0.194	89.3
0.781	0.107	107.1
0.391	0.408	125.5
0.195	0.709	142.9

Table 7.1: Measurement results. Data from http://zimmer.csufresno.edu/~da-vidz/Chem102/FluorideISE/FluorideISE.html.

From the plot in Figure 7.2, we can clearly identify the linear relationship between E versus log C with slope measured at -59.4 mV, which is very closed to the theoretical value -59.2 mV at 25 °C. This plot can give the concentration of any solution containing fluoride ion within the range of 0.195 mg/L and 200 mg/L by measuring the potential of the unknown solution.

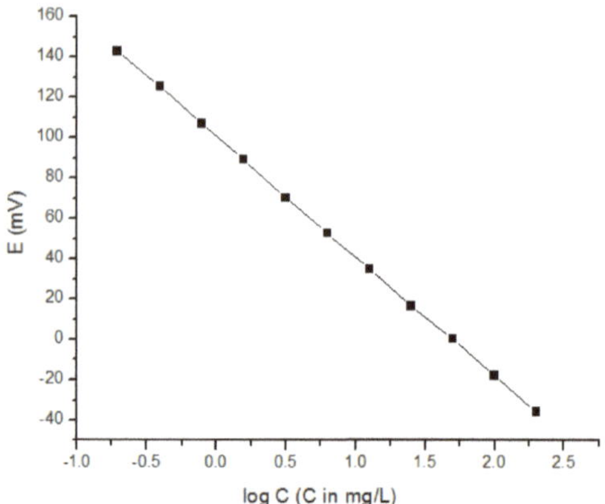

Figure 7.2: Plot of E versus log C. Data from http://zimmer.csufresno.edu/~davidz/Chem102/FluorideISE/FluorideISE.html.

Limitations of ISE

Though ISE is a cost-effective and useful technique, it has some drawbacks that cannot be avoided. The selective ion membrane only allows the measured ions to pass and thus the potential is only determined by this particular ion. However, the truth is there is no such membrane that only permits the passage of one ion, and so there are cases when there is more than one ion that can pass the membrane. As a result, the measured potential is affected by the passage of the unwanted ions. Also, because of its dependence on ion selective membrane, one ISE is only suitable for one ion and this may be inconvenient sometimes. Another problem worth noticing is that ion selective measures the concentration of ions in equilibrium at the surface of the membrane surface. This does matter much if the solution is dilute but at higher concentrations, the inter-ionic interactions between the ions in the solution tend to decrease the mobility of ions and thus the concentration near the membrane would be lower than that in the bulk. This is one source of inaccuracy of ISE. To better analyze the results of ISE, the user has to be aware of these inherent limitations of it.

Bibliography

R. Greef, R. Peat, L. M. Peter, D. Pletcher, and J. Robinson, *Instrumental Methods in Electrochemistry*, Ellis Horwood, Chichester (1985).

A. A. Khan and S. Shaheen, Determination of arsenate in water by anion selective membrane electrode using polyurethane–silica gel fibrous anion exchanger composite. *J. Hazard. Mater.*, 2014, **264**, 84.

J. E. O'Reilly, Determination of iodide in milk with an ion selective electrode. *J. Chem. Edu.*, 1979, **56**, 279.

D. S. Papastathopoulos and M. I. Karayannis, Construction and evaluation of a solid-state iodide selective electrode: A chemical instrumentation laboratory experiment. *J. Chem. Edu.*, 1980, **57**, 904.

F. Scholz, *Electroanalytical Methods: Guide to Experiments and Application*, 2nd edn., Springer, Berlin (2010).

G. Somer, U. T. Yilmaz, and Ş. Kalayc, Preparation and properties of a new solid state arsenate As(V) ion selective electrode and its application. *Tantala*, 2015, **142**, 120.

74

Chapter 8: X-ray Absorption Spectroscopy

Natalia Gonzalez Pech and Andrew R. Barron

Introduction

X-ray absorption spectroscopy (XAS) is a technique that uses synchrotron radiation to provide information about the electronic, structural, and magnetic properties of certain elements in materials. This information is obtained when X-rays are absorbed by an atom at energies near and above the core level binding energies of that atom. Therefore, a brief description about X-rays, synchrotron radiation and X-ray absorption is provided prior to a description of sample preparation for powdered materials.

X-rays and synchrotron radiation

X-rays were discovered by the Wilhelm Röntgen in 1895 (Figure 8.1). They are a form of electromagnetic radiation, in the same manner as visible light but with a very short wavelength, around 0.25 - 25 Å. As electromagnetic radiation, X-rays have a specific energy. The characteristic range is defined by soft versus hard X-rays. Soft X-rays cover the range from hundreds of eV to a few KeV, and the hard X-rays have an energy range from a few KeV up to around 100 KeV.

Figure 8.1: German physicist Wilhelm Conrad Röntgen (1845 1923) who received the first Nobel Prize in Physics in 1901 for the production and use of X-rays.

X-rays are commonly produced by X-ray tubes, where high-speed electrons strike a metal target. The electrons are accelerated by a high voltage towards the metal target; X-rays are produced when the electrons collide with the nuclei of the metal target.

Synchrotron radiation is generated when particles are moving at really high velocities and are deflected along a curved trajectory by a magnetic field. The charged particles are first accelerated by a linear accelerator (LINAC) (Figure 8.2); then, they are accelerated in a booster ring that injects the particles moving almost at the speed of light into the storage ring. There, the particles are accelerated toward the center of the ring each time their trajectory is changed so that they travel in a closed loop. X-rays with a broad spectrum of energies are generated and emitted tangential to the storage ring. Beamlines are placed tangential to the storage ring to use the intense X-ray beams at a wavelength that can be selected varying the set-up of the beamlines. Those are well suited for XAS measurements because the X-ray energies produced span 1000 eV or more as needed for an XAS spectrum.

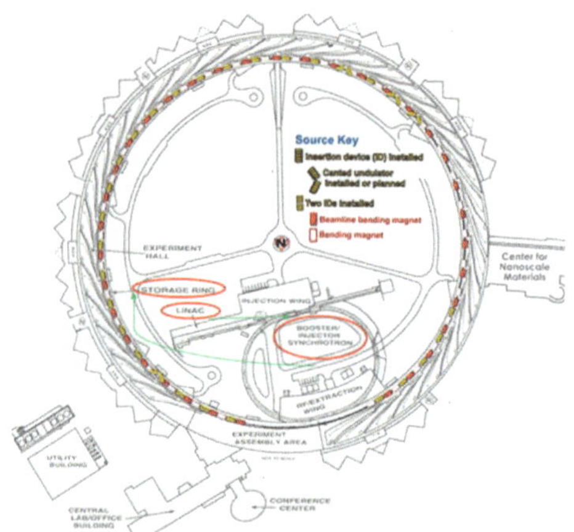

Figure 8.2: Scheme of a synchrotron and the particle trajectory inside it. Adapted from S. D. Kelly, D. Hesterberg, and B. Ravel in *Methods of Soil Analysis: Part 5, Mineralogical Methods*, Ed. A. L. Urely and R. Drees, Soil Science Society of America Book Series, Madison (2008).

X-ray absorption

Light is absorbed by matter through the photoelectric effect. It is observed when an X-ray photon is absorbed by an electron in a strongly bound core level (such as the 1s or 2p level) of an atom (Figure 8.3). In order for a particular electronic core level to participate in the absorption, the binding energy of this core level must be less than the energy of the incident X-ray. If the binding energy is greater than the energy of the X-ray, the bound electron will not be perturbed and will not absorb the X-ray. If the binding energy of the electron is less than that of the X-ray, the electron may be removed from its quantum level. In this case, the X-ray is absorbed and any energy in excess of the electronic binding energy is given as kinetic energy to a photoelectron that is ejected from the atom.

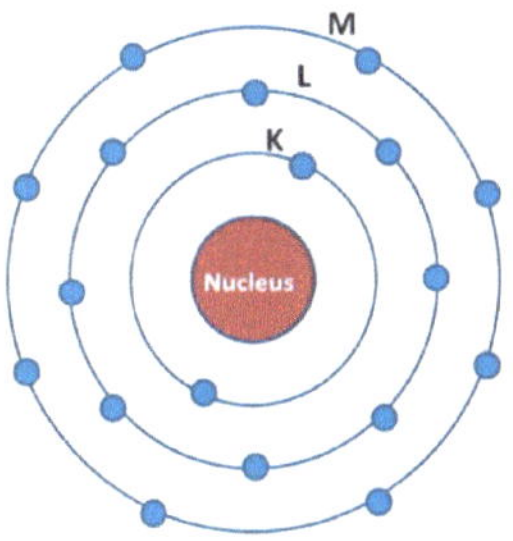

Figure 8.3: A schematic representation of the photoelectric effect when a photon with the right energy hits an electron, it is expelled.

When X-ray absorption is discussed, the primary concern is about the absorption coefficient, μ, which gives the probability that X-rays will be absorbed according to Beer's Law,

$$I = I_0 e^{-\mu t}$$

where I_0 is the X-ray intensity incident on a sample, t is the sample thickness, and I is the intensity transmitted through the sample.

The absorption coefficient, μ_E, is a smooth function of energy, with a value that depends on the sample density ρ, the atomic number Z, atomic mass A, and the X-ray energy E roughly as,

$$\mu_E \approx \frac{\rho Z^4}{AE^3}$$

When the incident X-ray has energy equal to that of the binding energy of a core-level electron, there is a sharp rise in absorption: an absorption edge corresponding to the promotion of this core level to the continuum. For XAS, the main concern is the intensity of μ, as a function of energy, near and at energies just above these absorption edges. An XAS measurement is simply a measure of the energy dependence of μ at and above the binding energy of a known core level of a known atomic species. Since every atom has core-level electrons with well-defined binding energies, the element to probe can be selected by tuning the X-ray energy to an appropriate absorption edge. These absorption edge energies are well-known. Because the element of interest is chosen in the experiment, XAS is element specific.

X-ray absorption fine structure

X-ray absorption fine structure (XAFS) spectroscopy, also named X-ray absorption spectroscopy, is a technique that can be applied for a wide variety of disciplines because the measurements can be performed on solids, gasses, or liquids, including moist or dry soils, glasses, films, membranes, suspensions or pastes, and aqueous solutions. Despites its broad adaptability with the kind of material used, there are samples which limits the quality of an XAFS spectrum. Because of that, the sample requirements and sample preparation are reviewed in this section as well the experiment design which are vital factors in the collection of good data for further analysis.

Experiment design

The main information can be obtained using XAFS spectra consist in small changes in the absorption coefficient (E), which can be measured directly in a transmission mode or indirectly using a fluorescence mode. Therefore, a good signal to noise ratio is required (better than 10^3). In order to obtain this signal to noise ratio, an intense beam is required (on the order 10^{10} photons/second or better), with the energy bandwidth of 1 eV or less, and the capability of scanning the energy of the incident beam over a range of about 1 KeV above the edge in a time range of seconds or few minutes. As a result, synchrotron radiation is preferred further than other kind of X-ray sources previously mentioned.

Beamline setup

Despite the setup of a synchrotron beamline is mostly done by the assistance of specialist beamline scientists, nevertheless, it is useful to understand the

system behind the measurement. The main components of a XAFS beamline, shown in Figure 8.4, are as follows:
- A harmonic rejection mirror to reduce the harmonic content of the X-ray beam.
- A monochromator to choose the X-ray energy.
- A series of slits which defines the X-ray profile.
- A sample positioning stage.

The detectors, which can be a single ionization detector or a group of detectors to measure the X-ray intensity.

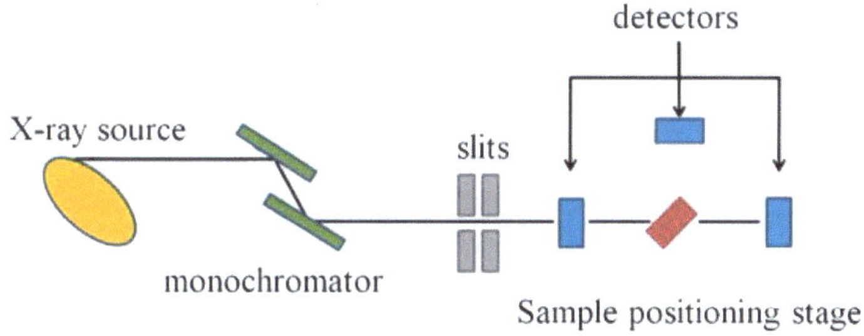

Figure 8.4: Schematic of the basic components of a XAFS beamline.

Slits are used to de ne the X-ray beam profile and to block unwanted X-rays. Slits can be used to increase the energy resolution of the X-ray incident on the sample at the expense of some loss in X-ray intensity. They are either fixed or adjustable slits. Fixed slits have a pre-cut opening of heights between 0.2 and 1.0 mm and a width of some centimeters. Adjustable slits use metal plates that move independently to de ne each edge of the X-ray beam.

Monochromator

The monochromator is used to select the X-ray energy incident on the sample. There are two main kinds of X-ray monochromators:
- The double-crystal monochromator, which consists of two parallel crystals.
- The channel-cut monochromator, which is a single crystal with a slot cut nearly through it.

Most monochromator crystals are made of silicon or germanium and are cut and polished such that a particular atomic plane of the crystal is parallel to the surface of the crystal as Si(111), Si(311), or Ge(111). The energy of X-rays diffracted by the crystal is controlled by rotating the crystals in the white beam.

Harmonic rejection mirrors

The harmonic X-ray intensity needs to be reduced, as these X-rays will adversely affect the XAS measurement. A common method for removing harmonic X-rays is using a harmonic rejection mirror. This mirror is usually made of Si for low energies, Rh for X-ray energies below the Rh absorption edge at 23 keV, or Pt for higher X-ray energies. The mirror is placed at a grazing angle in the beam such that the X-rays with fundamental energy are reflected toward the sample, while the harmonic X-rays are not.

Detectors

Most X-ray absorption measurements use ionization detectors. These contain two parallel plates separated by a gas-filled space that the X-rays travel through. Some of the X-rays ionize the gas particles. A voltage bias applied to the parallel plates separates the gas ions, creating a current. The applied voltage should give a linear detector response for a given change in the incident X-ray intensity. There are also other kinds as fluorescence and electron yield detectors.

Transmission and fluorescence modes

X-ray absorption measurements can be performed in several modes: transmission, fluorescence and electron yield; where the two first are the most common. The choice of the most appropriate mode to use in one experiment is a crucial decision.

The transmission mode is the most used because it only implies the measure of the X-ray flux before and after the beam passes the sample. Therefore, the adsorption coefficient is defined as,

$$\mu_E = \ln(I_0/I)$$

Transmission experiments are standard for hard X-rays, because the use of soft X-rays implies the use the samples thinner than 1 μm. Also, this mode should be used for concentrated samples. The sample should have the right thickness and be uniform and free of pinholes.

The fluorescence mode measures the incident flux I_0 and the fluorescence X-rays I_f that are emitted following the X-ray absorption event. Usually the fluorescent detector is placed at 90° to the incident beam in the horizontal plane, with the sample at an angle, commonly 45°, with respect to the beam, because in that position there is not interference generated because of the initial X-ray flux (I_0). The use of fluorescence mode is preferred for thicker samples or lower concentrations, even ppm concentrations or lower. For a highly concentrated sample, the fluorescence X-rays are reabsorbed by the absorber atoms in the sample, causing an attenuation of the fluorescence signal, it effect is named as self-absorption and is one of the most important concerns in the use of this mode.

Sample preparation for XAS

Sample requirements

Uniformity

The samples should have a uniform distribution of the absorber atom and have the correct absorption for the measurement. The X-ray beam typically probes a millimeter-size portion of the sample. This volume should be representative of the entire sample.

Thickness.

For transmission mode samples, the thickness of the sample is really important. It supposes to be a sample with a given thickness, t, where the total adsorption of the atoms is less than 2.5 adsorption lengths, $\mu Et \approx 2.5$; and the partial absorption due to the absorber atoms is around one absorption length $\Delta \mu Et \approx 1$, which corresponds to the step edge.

The thickness to give $\Delta \mu Et = 1$ is as,

$$t = \frac{1}{\Delta \mu} = \frac{1.66 \Sigma_i n_i M_i}{\rho \Sigma_i n_i [\sigma_i(E_+) - \sigma_i(E_-)]}$$

where ρ is the compound density, n is the elemental stoichiometry, M is the atomic mass, σE is the adsorption cross-section in barns/atom (1 barn = 10^{-24} cm^2) tabulated in McMaster tables, and E_+ and E_- are the just above and below the energy edge. This calculation can be accomplished using free download software.

Total X-ray adsorption.

For non-concentrate samples, the total X-ray adsorption of the sample is the most important. It should be related to the area concentration of the sample (ρt, in g/cm^2). The area concentration of the sample multiplied by the difference of the mass adsorption coefficient ($\Delta\mu E/\rho$) gives the edge step, where a desired value to obtain a good measure is an edge step equal to one, $(\Delta\mu E/\rho)\rho t \approx 1$.

The difference of the mass adsorption coefficient is given by,

$$\Delta\mu_E/\rho = \Sigma f_i[(\Delta\mu_E/\rho)_{i,(E+)} - (\Delta\mu_E/\rho)_{i,(E-)}]$$

where $(\mu_E/\rho)_i$ is the mass adsorption coefficient just above (E_+) and just below (E_-) of the edge energy and f_i is the mass fraction of the element *i*. Multiplying the area concentration, ρt, for the cross-sectional area of the sample holder, amount of sample needed is known.

Sample preparation

As was described in last section, there are diluted solid samples, which can be prepared onto big substrates or concentrate solid samples which have to be prepared in thin films. Both methods are following described. Liquid and gases samples can also be measured, but the preparation of those kind of sample is not discussed herein because it depends in the specific requirements of each sample. Several designs can be used as long they avoid the escape of the sample and the material used as container does not absorb radiation at the energies used for the measure.

Method 1

Step 1: The materials needed are showed in Figure 8.5: Kapton tape and film (Figure 8.6), a thin spatula, tweezers, scissors, weigh paper, mortar and pestle, and a sample holder. The sample holder can be made from several materials, as polypropylene, polycarbonate or Teflon.

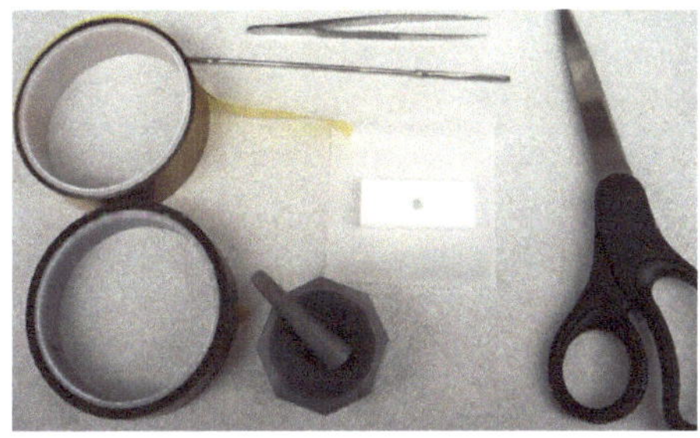

Figure 8.5: Several tools are needed for the sample preparation using Method 1.

Figure 8.6: The structure of poly-oxydiphenylene-pyromellitimide (Kapton).

Step 2: Two small squares of Kapton film are cut. One of them is placed onto the hole of the sample holder as shown Figure 8.7a. A piece of Kapton tape is placed onto the sample holder trying to minimize any air burble onto the surface and keeping the film as was previously placed Figure 8.7b. A side of the sample holder is now sealed in order to fill the hole (Figure 8.8).

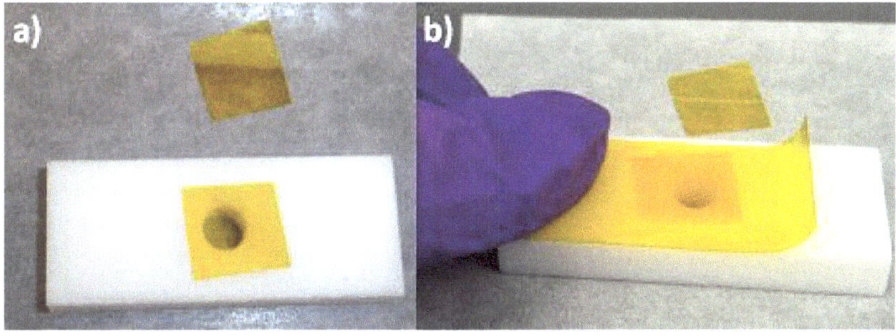

Figure 8.7: Preparing one face of the sample holder by (a) positioning a small piece of Kapton film onto the hole, which is held in place by Kapton tape (b).

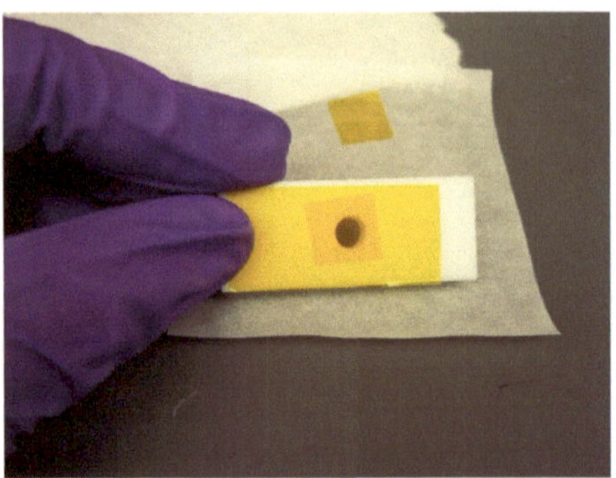

Figure 8.8: The side of the sample holder is closed.

Step 3: Before filling the sample holder, make sure your sample is a fine powder. Use the mortar to grind the sample (Figure 8.9).

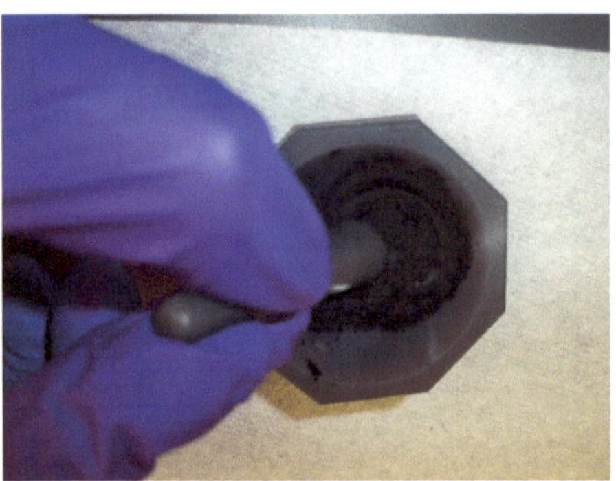

Figure 8.8: The sample is ground to be sure the grain size of the sample is homogeneous and small enough.

Step 4: Fill the hole with the powder. Make sure you have extra powder onto the hole (Figure 8.9a). With the spatula press the powder. The sample has to be as compact as possible (Figure 8.9b).

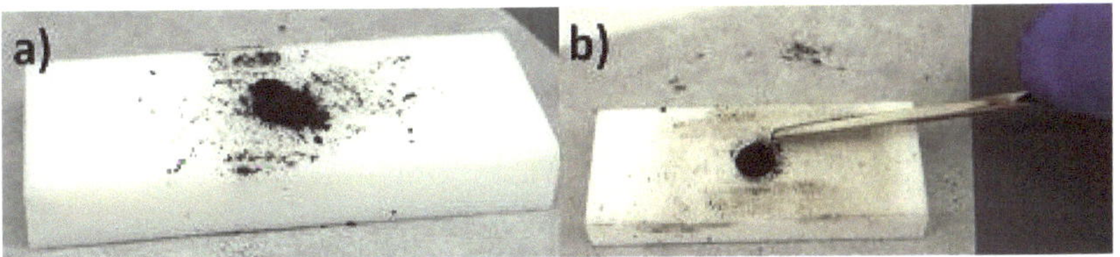

Figure 8.9: The sample holder is filled by (a) adding extra powder onto the hole then (b) compacting the sample with the spatula.

Step 5: Clean the surface of the slide. Repeat the Step 2. Your sample loaded in the sample holder should look as Figure 8.10.

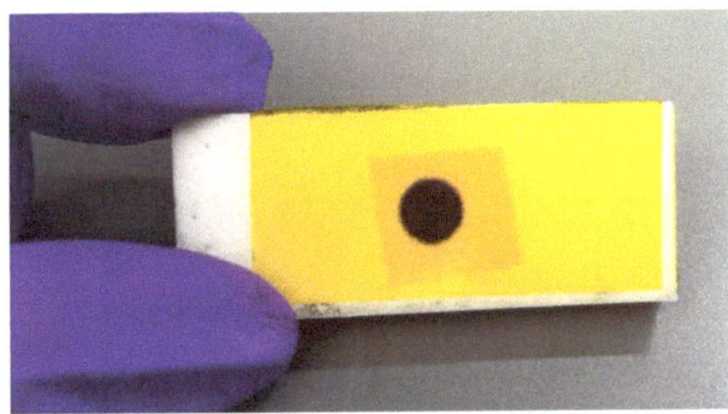

Figure 8.10: Sample loaded and sealed into the sample holder.

Method 2

Step 1: The materials needed are showed in Figure 8.11: Kapton tape, tweezers, scissors, weigh paper, mortar and pestle, tape and aluminum foil.

Figure 8.11: Several utensils are needed for the sample preparation using Method 2.

Step 2: Aluminum foil is placed as the work-area base. Kapton tape is place from one corner to the opposite one as shown Figure 8.12. Tape is put onto the extremes to fix it. In this case yellow tape was used in order to show where the tape should be placed but is better use Scotch invisible tape for the following steps.

Figure 8.12: Preparation of the work-area.

Step 3: The weigh paper is placed under the Kapton tape in one of the extremes. Sample is added onto that Kapton tape extreme (Figure 8.13). The function of the weigh paper is further recuperation of extra sample.

Figure 8.13: Add the sample onto an extreme of the Kapton tape.

Step 4: With one finger, the sample is dispersed along the Kapton tape, always in the same direction and taking care that the weigh paper is under the tape area is being used (Figure 8.14a). The finger should be slid several times making pressure in order to have a homogeneous and complete cover film (Figure 8.14b).

Figure 8.14: Making a thin film with a solid sample by (a) dispersing the solid along the Kapton tape and (b) repeated sliding several times to obtain a homogeneous film.

Step 5: The final sample covered Kapton tape should look like Figure 8.15. Cut the extremes in order to a further manipulation of the film.

Figure 8.15: A complete thin film.

Step 6: Using the tweezers, fold the film taking care that is well aligned and the fold is complete plane. Figure 8.16a shows the first folding, generating a 2 layers film. Figure 8.16b and Figure 8.16c shows the second and third folding, obtaining a 4- and 8-layer film. Sometimes a 4-layers film is good enough. You always can fold again to obtain bigger signal intensity.

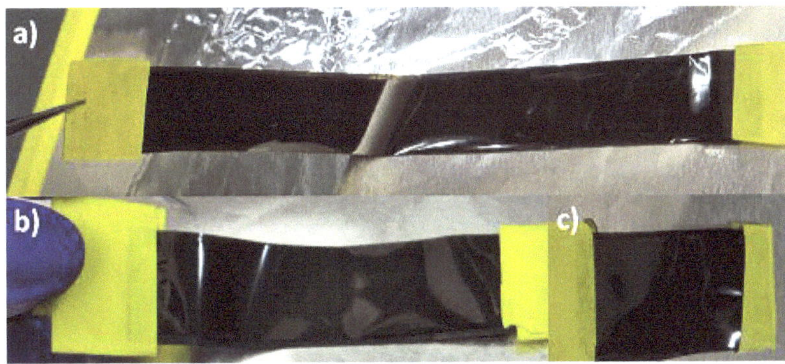

Figure 8.16: Folding of the thin film simple once results in a two-layer film (a) and after a second and third folding four- and eight-layers films are obtained (b and c, respectively).

Bibliography

B. D. Cullity and S. R. Stock. *Elements of X-ray Diffraction*, Prentice Hall, Upper Saddle River (2001).

F. Hippert, E. Geissler, J. L. Hodeau, E. Lelièvre-Berna, and J. R. Regnard. *Neutron and X-ray Spectroscopy,* Springer, Dordrecht (2006).

G. Bunker. *Introduction to XAFS: A practical guide to X-ray Absorption Fine Structure Spectroscopy*, Cambridge University Press, Cambridge (2010).

S. D. Kelly, D. Hesterberg, and B. Ravel in *Methods of Soil Analysis: Part 5, Mineralogical Methods*, Ed. A. L. Urely and R. Drees, Soil Science Society of America Book Series, Madison (2008).

Chapter 9: Neutron Activation Analysis

Jessica Heimann and Andrew R. Barron

Introduction

Neutron activation analysis (NAA) is a non-destructive analytical method commonly used to determine the identities and concentrations of elements within a variety of materials. Unlike many other analytical techniques, NAA is based on nuclear rather than electronic transitions. In NAA, samples are subjected to neutron radiation (i.e., bombarded with neutrons), which causes the elements in the sample to capture free neutrons and form radioactive isotopes, such as in

$$_{27}^{59}\text{Co} + {_0^1}\text{n} \rightarrow {_{27}^{60}}\text{Co}$$

The excited isotope undergoes nuclear decay and loses energy by emitting a series of particles that can include neutrons, protons, alpha particles, beta particles, and high-energy gamma ray photons. Each element on the periodic table has a unique emission and decay path that allows the identity and concentration of the element to be determined.

History

George de Hevesy (Figure 9.1) and Hilde Levi (Figure 9.2) published the first paper on the process of neutron activation analysis in 1936. They had discovered that rare earth elements such as dysprosium became radioactive after being activated by thermal neutrons from a radon-beryllium (^{266}Ra + Be) source. Using a Geiger counter to count the beta particles emitted, Hevesy and Levi were able to identify the rare earth elements by half-life. This discovery led to the increasingly popular process of inducing radioactivity and observing the resulting nuclear decay in order to identify an element, a process we now know as NAA.

Figure 9.1: Hungarian radiochemist George Charles de Hevesy (1885 - 1966) who received the Nobel Prize in Chemistry (1943) for the development of radioactive tracers.

Figure 9.2: German-Danish physicist Hilde Levi (1909 - 2003).

In the years immediately following Hevesy and Levi's discovery, however, the advancement of this technique was restricted by the lack of stable neutron sources and adequate spectrometry equipment. Even with the development of charged-particle accelerators in the 1930s, analyzing multi-element samples remained time-consuming and tedious. The method was improved in the mid-1940s with the availability of the X-10 reactor at the Oak Ridge National Laboratory (Figure 9.3), the first research-type nuclear reactor. As compared with the earlier neutron sources used, this reactor increased the sensitivity of NAA by a factor of a million. Yet the detection step of NAA still revolved around Geiger or proportional counters; thus, many technological advancements were still to come. As technology has progressed in the recent decades, the NAA method has grown tremendously, and scientists now have a plethora of neutron sources and detectors to choose from when analyzing a sample with NAA.

Figure 9.3: The X-10 graphite reactor at the Oak Ridge National Laboratory, in Oak Ridge, Tennessee.

Sample preparation

In order to analyze a material with NAA, a small sample of at least 50 milligrams must be obtained from the material, usually by drilling. It is suggested that two different samples are obtained from the material using two drill bits of different compositions. This will show any contamination from the drill bits and, thus, minimize error. Prior to irradiation, the small samples are encapsulated in vials of either quartz or high purity linear polyethylene.

Instrumentation

Neutron activation analysis works through the processes of neutron activation and radioactive decay. In neutron activation, radioactivity is induced by bombarding a sample with free neutrons from a neuron source. The target atomic nucleus captures a free neutron and, in turn, enters an excited state. This excited and therefore unstable isotope undergoes nuclear decay, a process in which the unstable nucleus emits a series of particles that can include neutrons, protons, alpha, and beta particles in an effort to return to a low-energy, stable state. As suggested by the several different particles of ionizing radiation listed above, there are many different types of nuclear decay possible. These are summarized in Figure 9.4.

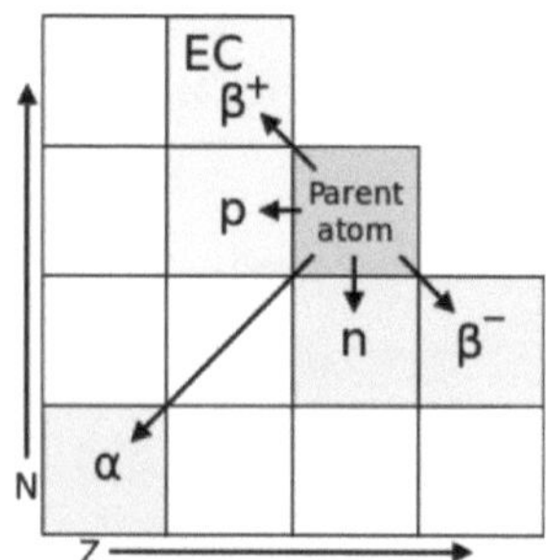

Figure 9.4: Transition diagram illustrating the changes in neutron number N and atomic number Z for different nuclear decay modes alpha decay (α), normal beta decay (β⁻), positron emission (β⁺), electron capture (EC), proton emission (p), and neutron emission (n). Permission to copy granted via the GNU Free Documentation License.

An additional type of nuclear decay is that of gamma radiation (denoted as γ), a process in which the excited nucleus emits high-energy gamma ray photons. There is no change in either neutron number N or atomic number Z, yet the nucleus undergoes a nuclear transformation involving the loss of energy. In order to distinguish the higher energy parent nucleus (prior to gamma decay) from the lower energy daughter nucleus (after gamma decay), the mass number of the parent nucleus is labeled with the letter m, which means metastable. An example of gamma radiation with the element technetium is provided in

$$^{99m}_{43}\text{Te} \rightarrow {}^{99}_{43}\text{Te} + {}^{0}_{0}\gamma$$

In NAA, the radioactive nuclei in the sample undergo both gamma and particle nuclear decay. Figure 9.5 presents a schematic example of nuclear decay. After capturing a free neutron, the excited ^{60m}Co nucleus undergoes an internal transformation by emitting gamma rays. The lower-energy daughter nucleus ^{60}Co, which is still radioactive, then emits a beta particle. This results in a high-energy ^{60}Ni nucleus, which once again undergoes an internal transformation by emitting gamma rays. The nucleus then reaches the stable ^{60}Ni state.

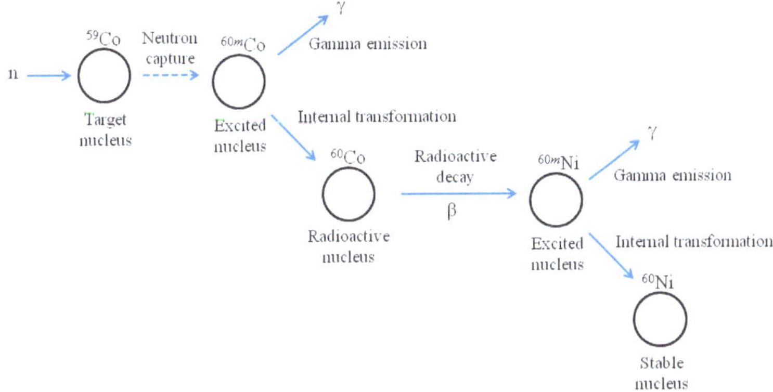

Figure 9.5: Scheme of neutron activation analysis with ^{59}Co as the target nucleus.

Although alpha and beta particle detectors do exist, most detectors used in NAA are designed to detect the gamma rays that are emitted from the excited nuclei following neutron capture. Each element has a unique radioactive emission and decay path that is scientifically known. Thus, based on the path and the spectrum produced by the instrument, NAA can determine the identity and concentration of the element.

Neutron sources

As mentioned above, there are many different neutron sources that can be used in modern-day NAA. A chart comparing three common sources is shown in Table 9.1.

Gamma and particle detectors

As mentioned earlier, most detectors used in NAA are designed to detect the gamma rays emitted from the decaying nucleus. Two widely used gamma detectors are the scintillation type and the semiconductor type. The former uses a sensitive crystal, often sodium iodide that is doped with thallium (NaI(Tl)), that emits light when gamma rays strike it. Semiconductor detectors, on the other hand, use germanium to form a diode that produces a signal in response to gamma radiation. The signal produced is proportional to the energy of the emitted gamma radiation. Both types of gamma detectors have excellent sensitivity with detection limits ranging from 0.1 to 10^6 nanogram element per gram sample, but semiconductor type detectors usually have superior resolution.

Source type	Description	Example(s)	Typical output
Isotopic neutron sources	Certain isotopes undergo spontaneous fission and release neutrons as they decay.	^{226}Ra(Be), ^{124}Sb(Be), ^{241}Am(Be), ^{252}Cf	10^5 - 10^7 s^{-1} GBq^{-1} or 2.2×10^{12} s^{-1} g^{-1} for ^{252}Cf
Particle accelerators or neutron generators	Particle accelerators produce neutrons by colliding hydrogen, deuterium, and tritium with target nuclei such as deuterium, tritium, lithium, and beryllium.	Acceleration of deuterium ions toward a target containing deuterium or tritium, resulting in the reactions ^{2}H(^{2}H,n)^{3}He and ^{3}H(^{2}H,n)^{4}He	10^8 - 10^{10} s^{-1} for the first deuterium on deuterium reactions and 10^9 - 10^{11} s^{-1} for deuterium on tritium reactions
Nuclear research reactors	Within nuclear reactors, large atomic nuclei absorb neutrons and undergo nuclear fission. The nuclei split into lighter nuclei, which releases energy, radiation, and free neutrons.	^{235}U and ^{239}Pu	10^{15} - 10^{18} m^{-2} s^{-1}

Table 9.1: Different neutron sources.

Furthermore, particles detectors designed to detect the alpha and beta particles that are emitted in nuclear decay are also available; however, gamma detectors are favorable. Particle detectors require a high vacuum since atmospheric gases in the air can absorb and affect the emission of these particles. Gamma rays are not affected in this way.

Variations/parameters

INAA versus RNAA

Instrumental neutron activation analysis (INAA) is the simplest and most widely used form of NAA. It involves the direct irradiation of the sample, meaning that the sample does not undergo any chemical separation or treatment prior to detection. INAA can only be used if the activity of the other radioactive isotopes in the sample does not interfere with the measurement of the element(s) of interest. Interference often occurs when the element(s) of interest are present in trace or ultra-trace amounts. If interference does occur, the activity of the other radioactive isotopes must be removed or eliminated. Radiochemical separation is one way to do this. NAA that involves sample decomposition and elemental separation is known as radiochemical neutron activation analysis (RNAA). In RNAA, the interfering elements are separated from the element(s) of interest through an appropriate separation method. Such methods include extractions, precipitations, distillations, and ion exchanges. Inactive elements and matrices are often added to ensure appropriate conditions and typical behavior for the element(s) of interest. A schematic comparison of INAA and RNAA is shown in Figure 9.6.

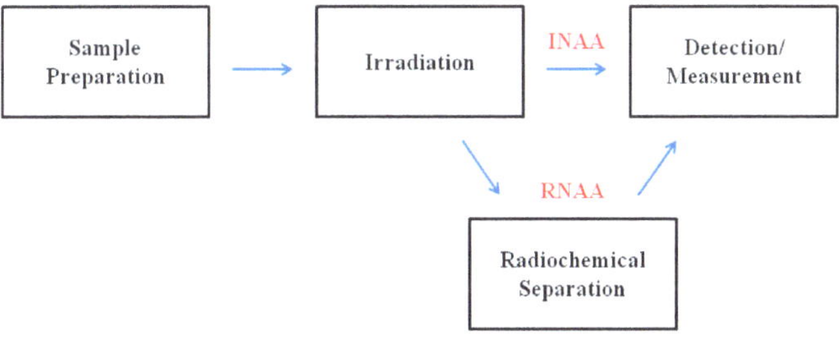

Figure 9.6: Schematic Comparison of INAA and RNAA.

ENAA versus FNAA

Another experimental parameter that must be considered is the kinetic energy of the neutrons used for irradiation. In epithermal neutron activation analysis (ENAA), the neutrons known as epithermal neutrons are partially moderated in the reactor and have kinetic energies between 0.5 eV to 0.5 MeV. These are lower-energy neutrons as compared to fast neutrons, which are used in fast neutron activation analysis (FNAA). Fast neutrons are high-energy, unmoderated neutrons with kinetic energies above 0.5 MeV.

PGNAA versus DGNAA

The final parameter to be discussed is the time of measurement. The nuclear decay products can be measured either during or after neutron irradiation. If the gamma rays are measured during irradiation, the procedure is known as prompt gamma neutron activation analysis (PGNAA). This is a special type of NAA that requires additional equipment including an adjacent gamma detector and a neutron beam guide. PGNAA is often used for elements with rapid decay rates, elements with weak gamma emission intensities, and elements that cannot easily be determined by delayed gamma neutron activation analysis (DGNAA) such as hydrogen, boron, and carbon. In DGNAA, the emitted gamma rays are measured after irradiation. DGNAA procedures include much longer irradiation and decay periods than PGNAA, often extending into days or weeks. This means that DGNAA is ideal for long-lasting radioactive isotopes. A schematic comparison of PGNAA and DGNAA is shown below in Figure 9.7.

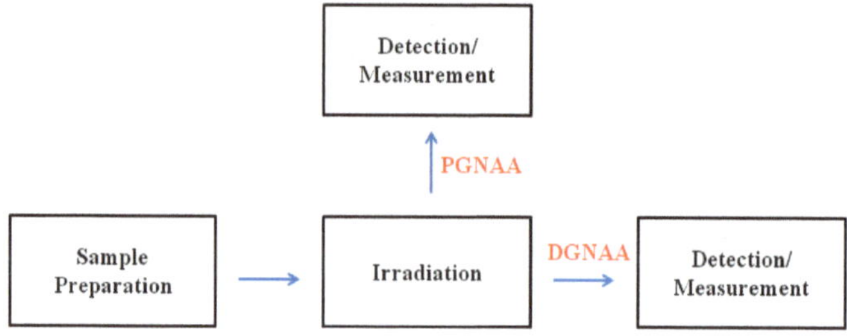

Figure 9.7: Schematic Comparison of PGNAA and DGNAA.

Examples

Characterizing archaeological materials

Throughout recent decades, NAA has often been used to characterize many different types of samples including archaeological materials. In 1961, the Demokritos nuclear reactor, a water moderated and cooled reactor, went critical at low power at the National Center for Scientific Research Demokritos (NCSR Demokritos) in Athens, Greece (Figure 9.8). Since then, NCSR Demokritos has been a leading center for the analysis of archaeological materials. Ceramics, carbonates, silicates, and steatite (soapstone) are routinely analyzed at NCSR Demokritos with NAA.

Figure 9.8. The National Center for Scientific Research Demokritos (NCSR Demokritos) in Athens, Greece.

A routine analysis begins by weighing and placing 130 milligrams of the powdered sample into a polyethylene vial. Two batches of ten vials, eight samples and two standards, are then irradiated in the Demokritos nuclear reactor for 45 minutes at a thermal neutron flux of 6×10^{13} neutrons cm^{-2}.s^{-1}. The first measurement occurs seven days after irradiation. The gamma ray emissions of both the samples and standards are counted with a germanium gamma detector (semiconductor type) for one hour. This measurement determines the concentrations of the following elements: As, Ca, K, La, Lu, Na, Sb, Sm, U, and Yb. A second measurement is performed three weeks after irradiation in which the samples and standards are counted for two hours. In this measurement, the concentrations of the following elements are determined: Ba, Ce, Co, Cr, Cs, Eu, Fe, Hf, Nd, Ni, Rb, Sc, Ta, Tb, Th, Zn, and Zr.

Using the method described above, NCSR Demokritos analyzed 195 samples of black-on-red painted pottery from the late Neolithic age in what is now known as the Black-On-Red Pottery Project. An example of black-on-red painted pottery is shown in Figure 9.9.

Figure 9.9: An example of black-on-red painted pottery from the late Neolithic age.

This project aimed to identify production patterns in this ceramic group and explore the degree of standardization, localization, and scale of production from 14 sites throughout the Strymonas Valley in northern Greece (Figure 9.10). NCSR Demokritos also sought to analyze the variations in pottery traditions by differentiating so-called ceramic recipes. By using NAA, NCSR Demokritos was able to determine the unique chemical make-ups of the many pottery fragments. The chemical patterning revealed through the analyses suggested that the 195 samples of black- on-red Neolithic pottery came from four distinct productions areas with the primary production area located in the valley of the Strymon and Angitis rivers. Although distinct, the pottery from the four different geographical areas all had common technological and stylistic characteristics, which suggests that a level of standardization did exist throughout the area of interest during the late Neolithic age.

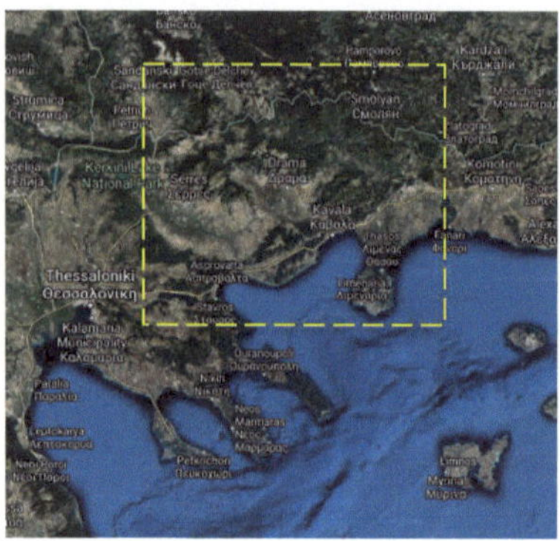

Figure 9.10: Location of the Strymonas Valley in northern Greece.

Determining elemental concentrations in blood

Additionally, NAA has been used in hematology laboratories to determine specific elemental concentrations in blood and provide information to aid in the diagnosis and treatment of patients. Identifying abnormalities and unusual concentrations of certain elements in the bloodstream can also aid in the prediction of damage to the organ systems of the human body.

NAA has been used to determine the concentrations of sodium and chlorine in blood serum. In order to investigate the accuracy of the technique in this setting, 26 blood samples of healthy male and female donors aged between 25 and 60 years and weighing between 50 and 85 kilograms were selected from the Paulista Blood Bank in São Paulo. The samples were initially irradiated for 2 minutes at a neutron flux ranging from approximately 1×10^{11} - 6×10^{11} neutrons cm^{-2}.s^{-1} and counted for 10 minutes using a gold activation detector. The procedure was later repeated using a longer irradiation time of 10 minutes. The determined concentrations of sodium and chlorine were then compared to standard values. The NAA analyses resulted in concentrations that strongly agreed with the adopted reference value. For example, the chlorine concentration was found to be 3.41 - 3.68 µg/µL of blood, which correlates closely to the reference value of 3.44 - 3.76 µg/µL of blood. This illustrates that NAA can accurately measure elemental concentrations in a variety of materials including blood samples.

Limitations

Although NAA is an accurate (~5%) and precise (<0.1%) multi-element analytical technique, it has several limitations that should be addressed. Firstly, samples irradiated in NAA will remain radioactive for a period of time (often years) following the analysis procedures. These radioactive samples require special handling and disposal protocols. Secondly, the number of the available nuclear reactors has declined in recent years. In the United States, only 31 nuclear research and test reactors are currently licensed and operating. A map of these reactors is provided in Figure 9.11.

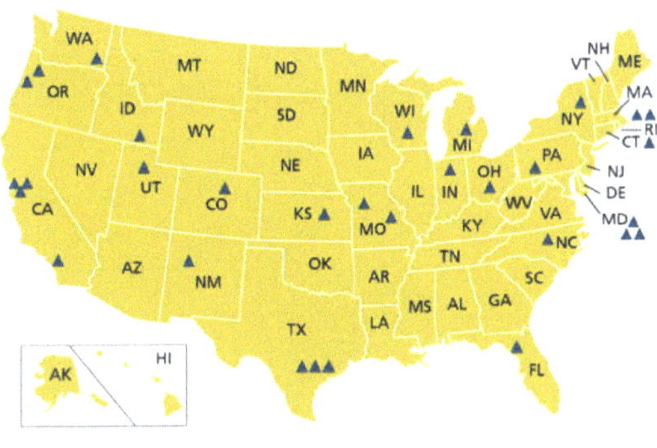

Figure 9.11: Map of US nuclear research and test reactors. Reproduced from *Map of Research and Test Reactor Sites,* http://www.nrc.gov/reactors/operating/map-nonpower-reactors.html.

As a result of the declining number of reactors and irradiation facilities in the nation, the cost of neutron activation analysis has increased. The popularity of NAA has declined in recent decades due to both the increasing cost and the development of other successful multi-element analytical methods such as inductively coupled plasma atomic emission spectroscopy (ICP-AES).

Bibliography

Z. B. Alfassi, *Activation Analysis*, CRC Press, Boca Raton (1990).

P. Bode, A. Byrne, Z. Chai, A. Chatt, V. Dimic, T. Z. Hossain, J. Ku£era, G. C. Lalor, and R. Parthasarathy, *Report of an Advisory Group Meeting Held in Vienna*, 22-26 June 1998, IAEA, Vienna, 2001, 1.

V. P. Guinn, New essential trace elements for the life sciences. *Biol. Trace Elem. Res.*, 1990, **26-27**, 1.

L. Hamidatou, H. Slamene, T. Akhal, and B. Zouranen, in *Interdisciplinary Research Fundamentals and Cutting Edge Applications*, Ed. F. Khar, InTech, Rijeka (2013).

V. Kilikoglou, A. P. Grimanis, A. Tsolakidou, A. Hein, D. Malalmidou, and Z. Tsirtsoni, Neutron activation patterning of archaeological materials at the National Center for Scientific Research 'Demokritos': the case of black-on-red neolithic pottery from Macedonia, Greece. *Archaeometry*, 2007, **49**, 301.

S. S. Nargolwalla and E. P. Przybylowicz, Activation analysis with neutron generators, *Nucl. Technol.*, 1975, **25**, 166.

M. Pollard and C. Heron, *Archaeological Chemistry*, Royal Society of Chemistry, Cambridge (1996).

B. Zamboi, L. C. Oliveira, and L. Dalaqua Jr., Diagnostic application of absolute neutron activation analysis in hematology. *Americas Nuclear Energy Symposium*, Miami (2004).

Chapter 10: Total Carbon Analysis

Julian Cooper and Andrew R. Barron

Introduction

Carbon is one of the more abundant elements on the planet; all living things and many non-living things have some form of carbon in them. Having the ability to measure and characterize the carbon content of a sample is of extreme value in a variety of different industries and research environments.

Total carbon (TC) content is just one important piece of information that is needed by analysts concerned with the carbon content of a sample. Having the knowledge of the origin of carbon in the sample, whether it be derived from organic or inorganic material, is also of extreme importance. For example, oil companies are interested in finding petroleum, a carbon containing material derived from organic matter, knowing the carbon content and the type of carbon in a sample of interest can mean the difference between investing millions of dollars and not doing so. Regulatory agencies like the U.S. Environmental Protection Agency (EPA) is another such example, where regulation of the carbon content and character of that carbon is essential for environmental and human health.

Considering the importance of identifying and quantifying the carbon content of an analyte, it may be surprising to learn that there is no one method to measure the carbon content of a sample. Unlike other techniques, no fancy instrument is required (although some exists that can be useful). In fact, methods to measure the different forms of carbon (organic or inorganic) are different themselves because they take advantage of the different properties characteristics to the carbon content you are measuring, in fact you will most likely use multiple techniques to fully characterize the carbon content of a sample, not just one.

Measurements of carbon content are related, and therefore measurement of either total carbon content (TC), total inorganic carbon content (TIC) and total organic carbon content (TOC) is related to the other two by,

$$TC = TIC + TOC$$

This means that measurement of two variables can indirectly give you the third, as there are only two classes of carbon: organic carbon and inorganic carbon.

Herein several of the methods used in measuring the TOC, TIC and TC for samples will be outlined. Not all samples require the same kind of instruments and methods. The goal of this module is to get the reader to see the simplicity of some of these methods and understand the need for such quantification and analysis.

Measurement of total organic carbon (TOC)

Sample and sample preparation

The total organic carbon content for a variety of different samples can be determined; there are very few samples that cannot be measured for total carbon content. Before treatment, a sample must be homogenized, whereby a sample is mixed or broken such that a measurement done on the sample can be representative of the entire sample. For example, if our sample were a rock, we would want to make sure that the inner core of the rock, which could have a different composition than the outer surface, were being measured as well. Not homogenizing the sample would lead to inconsistent and perhaps irreproducible results. Techniques for homogenization vary wildly, depending on the sample, different techniques exist.

Dissolution of total inorganic carbon

In order to measure the organic carbon content in a sample, the inorganic sources of carbon, which exist in the form of carbonate and bicarbonate salts and minerals, must be removed from the sample. This is typically done by treating the sample with non-oxidative acids such as H_2SO_4 and HCl, releasing CO_2 and H_2O,

$$2HCl + CaCO_3 \rightarrow CaCl_2 + CO_2 + H_2O$$

$$HCl + NaHCO_3 \rightarrow NaCl + CO_2 + H_2O$$

Non oxidative acids are chosen such that minimal amounts of organic carbon are affected. Although the selection of acid chosen to remove the inorganic sources of carbon is important; depending on your measurement technique, acids may interfere with the measurement. For example, in the wet measurement technique that will be discussed later, the counter ion Cl⁻ will add systematic error to the measurement.

Treatment of a sample with acid is intended to dissolve all inorganic forms of carbon in the sample. In selectively digesting and dissolving inorganic forms of carbon, be it aqueous carbonates or bicarbonates or trapped CO_2, one can selectively remove inorganic sources of carbon from organic ones; thereby leaving behind, in theory, only organic carbon in the sample.

It becomes apparent, in this treatment, the importance of sample homogenization. Using the rock example again. If a rock is treated with acid without homogenizing, the inorganic carbon at the surface of the sample may be dissolved. Only with homogenization can the acid dissolve in inorganic carbon on the inside of the rock. Otherwise this inorganic carbon may be interpreted as organic carbon, leading to gross errors in total organic carbon determination.

Shortcomings in the dissolution of inorganic carbon

A large problem and a potential source of error in technique measurement are the assumptions that have to be made, particularly in the case of TOC measurement, that all of the inorganic carbon has been washed away and separated from the sample. There is no way to distinguish TOC or TIC spectroscopically, the experimenter is forced to assume that they are looking at is all organic carbon or all inorganic carbon, when in reality there may be some of both still on the sample.

Quantitative measurement of TOC

Most TOC quantification methods are destructive in nature. The destructive nature of the methods means that none of the sample may be recovered. Of the methods, there are two destructive techniques that will be discussed in this module. The first is the wet method to measure TOC of solid sediment samples, and the second is dry combustion.

Wet methods

Sample preparation

Following sample pre-treatment with inorganic acids to dissolve away any inorganic material from the sample, a known amount of potassium dichromate ($K_2Cr_2O_7$, Figure 10.1) in concentrated sulfuric acid are added to the sample as per the Walkey-Black procedure, a well-known wet technique.

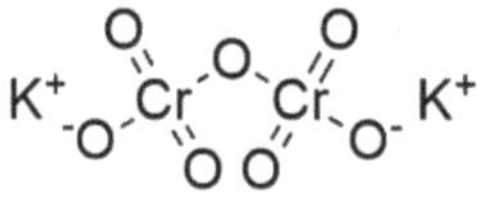

Figure 10.1: The structure of potassium dichromate ($K_2Cr_2O_7$).

The amount of dichromate and H_2SO_4 added can vary depending on the expected organic carbon content of the sample, typically enough H_2SO_4 is added such that the solid potassium dichromate dissolves in solution. The mixture of potassium dichromate with H_2SO_4 is an exothermic one, meaning that heat is evolved from the solution. As the dichromate reacts according to,

$$2Cr_2O_7^{2-} + 3C^0 + 16H^+ \rightarrow 4Cr^{3+} + 3CO_2 + 8H_2O$$

The solution will bubble away CO_2. Because the only source of carbon in the sample is in theory the organic forms of carbon (assuming adequate pre-treatment of the sample to remove the inorganic forms of carbon), the evolved CO_2 comes from organic sources of carbon.

Elemental forms of carbon in this method present problems for oxidation of elemental carbon to CO_2, meaning that not all of the carbon will be converted to CO_2, which will lead to an underestimation of total organic carbon content in the quantification steps. In order to facilitate the oxidation of elemental carbon, the digestive solution of dichromate and H_2SO_4 is heated at 150 °C for some time (~30 min, depending on total carbon content in the sample and the amount of dichromate added). It is important that the solution not be heated above 150 °C, as decomposition of the dichromate solution.

Other shortcomings, in addition to incomplete digestion, exist with this method. Fe^{2+} and Cl^- in the sample can interfere with the chromate solution, Fe^{2+} can be oxidized to Fe^{3+} and Cl^- can form CrO_2Cl_2 leading to systematic error towards higher organic carbon content. Conversely MnO_2, like dichromate, will oxidize organic carbon, thereby leading to a negative bias and an underestimation of TOC content in samples.

In order to counteract these biases, several additives can be used in the pre-treatment process. Fe^{2+} can be oxidized with mild oxidant phosphoric acid, which will not oxidize organic carbon. Treatment of the digestive solution with $AgSO_2$ can precipitate silver chloride. MnO_2 interferences can be dealt with using $FeSO_4$, where the oxidation power of the manganese is dealt with by taking the iron(II) sulfate to the +3 oxidation state. Any excess iron(II) can be dealt with using phosphoric acid.

Quantification of TOC

What follows sample treatment, where all of the organic carbon has been digested, is a titration to oxidize the excess dichromate in the sample. Comparing the excess that is titrated to the amount that was originally added to the original solution, one can do stoichiometric calculations and calculate the amount of dichromate that oxidized the organic carbon in the sample, thereby allowing the determination of TOC in the sample. How this titration is run is up to the user. Manual, potentiometric, titrations are all available to the investigator doing the TOC measurement, as well as some others.

Manual titrations are similar to any other type of manual titration method. An indicator must be used in manual titrations, and in the case of this wet method, commercially available ferroin is used. Titrant is typically ferrous ammonium sulfate. Titrant is added until equivalence is reached. Indicative of reaching equivalence is color change catalyzed by the indicator. Depending on the sample measured color change may be difficult to notice.

Insertion of platinum electrodes to the sample can be used to measure conductance of sample using potentiometric titration. When sample reached endpoint, conductance will essentially be zero or whatever the endpoint of the solution was set to. This method presents several advantages over manual titration methods because titration can be automated to respond to feedback from platinum electrodes, so equivalence point determination is not color dependent.

Alternative to titration methods, capture of evolved CO_2 presents another feasible quantification method, as oxidized organic carbon will be evolved as CO_2. Carbon dioxide can be captured on absorbent material such as ascarite

(NaOH) or other tared (unladed weight) absorbent, whose mass change as a result of absorbed CO_2 can be measured, or the absorbed CO_2 could be desorbed and quantified via IR non-dispersive cell.

Disadvantages of wet technique

Measurement of TOC via the described wet techniques is a rather crude method to measure organic carbon content in a sample. The technique relies on several assumptions that in reality are not wholly accurate, leading to TOC values that are in reality an approximate.

The treatment with acid to remove the inorganic forms of carbon assumes that all of the inorganic carbon is removed and washed away in the acid treatment, but in reality, this is probably not true, as some inorganic carbon will cling to the sample and be quantified incorrectly.

In the digestion process, which assumes that all of the carbon in the sample which is already presumed to be entirely organic carbon is completely converted carbon dioxide, taking no account for the possible solubility of the carbon dioxide in the wet sample or incomplete oxidation of carbon in the sample.

The wet method to measure TOC relies on the use of dichromate, while a very good oxidant, is a very toxic reagent with which to analysis.

TOC measurement of water

As mentioned previously, measurement of TOC levels in water is extremely valuable to regulatory agencies concerned with water quality. The presence of organic carbon in a substance that should have no carbon is of concern. Measurement of TOC in water uses a variant of the wet method in order to avoid highly toxic oxidants: typically, a persulfate salt (Figure 10.2) is used as an oxidant instead of dichromate.

Figure 10.2: Structural representation of persulfate salt, in this case the potassium salt. Breaking of oxygen-oxygen bond responsible for radical-induced oxidation.

The procedure for measuring TOC levels in water is essentially the same as in the typical wet oxidation technique. The water is first acidified to remove inorganic sources of carbon. Now because water is being measured, one cannot simply wash away the inorganic carbon. The inorganic carbon escapes from the water solution as CO_2. The remaining carbon in the solution is thought to be organic. Treatment of the solution with persulfate will do nothing. Irradiation of the solution treated with persulfate with UV radiation or heating will activate a radical species. This radical species will mediate oxidation of the organic carbon to CO_2, which can then be quantified by similar methods as the traditional wet oxidation technique.

Dry methods

As an alternative to technique for TOC measurement, dry techniques present several advantages over wet techniques. Dry techniques frequently involve the measurement of evolved carbon from the combustion of a sample. In this section of the module, TOC measurements using dry techniques will be discussed.

Sample pre-treatment

Like in the wet-oxidation case, measurement of TOC by dry techniques requires the removal of inorganic forms of carbon, and therefore samples are treated with inorganic acids to do so. The inorganic acids are washed away, and theoretically only organic forms of carbon remain. Before combustion of the sample, the treated sample must be completely dried so as to remove any moisture from the sample. In the case where non-volatile organics are present, or where little concern about the escape of organic material exists (e.g., rock samples or Kerogen), sample can be placed in a 100 °C oven overnight. In the case where evolution of organic matter at slightly elevated temperatures is a problem, drying can be performed under vacuum or in the presence of drying agent. Volatile organics are di cult to measure using dry techniques because the sample needs to be without moisture, and removal of moisture by any technique will most likely remove volatile organics.

Sample quantification

As mentioned before, quantification of TOC in the dry quantification method will proceed via complete combustion of the sample in a carbon free atmosphere (typically a pure oxygen atmosphere). Quantification of sample is performed via non-dispersive infrared detection cell. A characteristic asymmetric stretching at 2350 cm^{-1} can be seen for CO_2. The intensity of this infrared signal CO_2 is proportional to the quantity of CO_2 in the sample. Therefore, in order to translate signal intensity to amount, a calibration curve is constructed from known amounts of pure calcium carbonate, looking specifically at the intensity of the CO_2 peak. One may point out that calcium carbonate is an inorganic source of carbon, but it is important to note that the source of carbon has no effect on its quantification. Preparation of a calibration curve follows similar preparation as to an analyte, while no pre-treatment with acid is needed, the standards must be thoroughly dried in an oven. When a sample is ready to be analyzed, it is first weighed on some form of analytical balance, and then placed in the combustion analyzer, such as a LECO analyzer, where the oven and the non-dispersive IR cell are one machine (Figure 10.3).

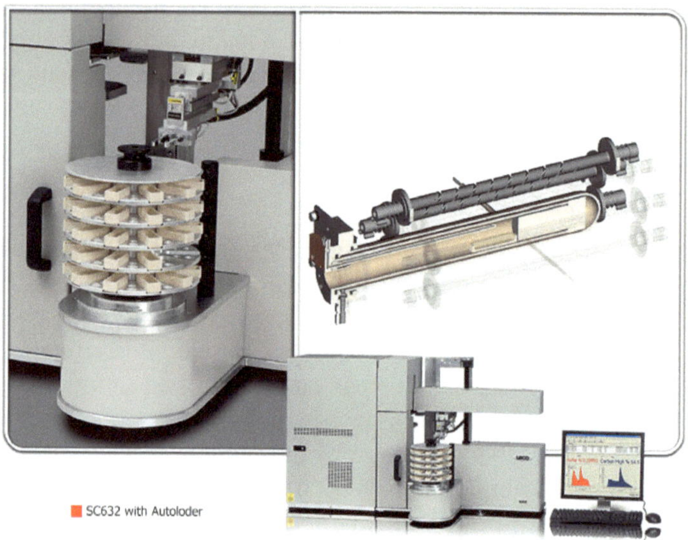

Figure 10.3: Example of a LECO analyzer. Samples placed in small cream-colored trays and combusted in oven under oxygen atmosphere (big box lower right) computer output to show carbon content. Material used with permission by LECO Corporation

Combustion proceeds at temperatures in the excess of 1350 °C in a stream of pure oxygen. Comparing the intensity of your characteristic IR peak to the intensities of the characteristic IR peaks of your known standards, the TOC of the sample can be determined. By comparing the mass of the sample to the mass of carbon obtained from the analyzer, the % organic carbon in the sample can be determined according to

$$\%TOC = \text{mass carbon/mass sample}$$

Use of this dry technique is most common for rock and other solid samples. In the oil and gas industry, it is extremely important to know the organic carbon content of rock samples in order to ascertain production viability of a well. The sample can be loaded in the LECO combustion analyzer and pyrolyzed in order to quantify TOC.

Measurement of total carbon (TC)

The total carbon in a sample (TC) is the sum of the inorganic forms of carbon and organic forms of carbon in a sample. It is known that no other sources of carbon contribute to the TC determination because no other sources of carbon exist. So in theory, if one could quantify the TOC by a method described in the previous section, and follow that with a measurement of the TIC in the pre-treatment acid waste, one could find the TC of a sample by summing the value obtained for TIC and the value obtained for TOC. However, in TC quantification this is hardly done: partly in order to avoid propagation of error associated with the other two methods, also cost restraints.

In measuring TC of a sample, the same dry technique of combustion of the sample is used, just like in the quantification of TOC. The same analyzer used to measure TOC can handle a TC measurement. No sample pre-treatment with acid is needed, so it is important to remember that the characteristic peak of CO_2 now seen is representative of the carbon of the entire sample. Now, the TIC carbon of the sample can be found as well. Subtraction of the TOC from the measured TC in the analyzer gives the value for TIC.

Measurement of total inorganic carbon (TIC)

Direct methods to measure the TIC of a sample, in addition to indirect measurement are possible. Typical TIC measurements are done on water samples, where the alkalinity and hardness of water is a result of inorganic carbonates, be it bicarbonate or carbonate. Treatment of these types of samples follows similar procedures to treatment of samples for organic carbon. A sample of water is acidified, such that the equilibrium,

$$CO_2 + H_2O \rightleftharpoons H_2CO_3 \rightleftharpoons HCO_3^- + H^+$$

obeys Le Chatelier's principle and favors the release of CO_2.

The CO_2 released can be measured in a variety of different ways. As with the combustion technique for measuring TC and TOC, measurement of the intensity of the characteristic IR stretch for CO_2 compared to standards can be used to quantity of TIC in a sample. However, in this case, it is emission of IR radiation that is measured, not absorption. An instrument that can do such a measurement is a FIRE-TIC, meaning Flame IR emission. This instrument consists of a purge like devices connected to a FIRE detector, as shown in Figure 10.4.

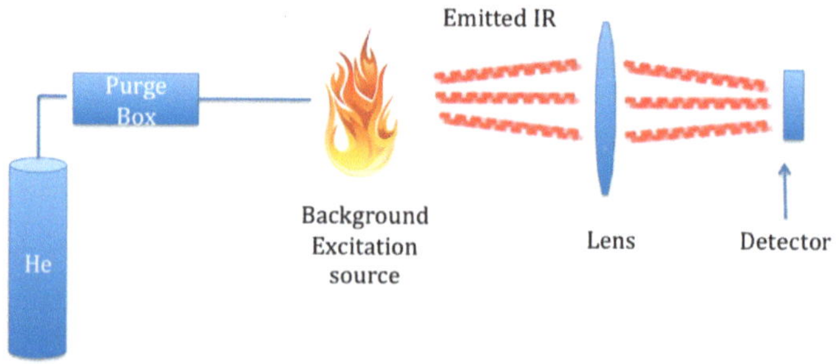

Figure 10.4: FIRE-TIC instrument. Sample is placed in a degassing purge box, in which typically helium or other IR inactive gas. As the gas passes through the sample CO_2 is released from sample.

Summary

Measurement of Carbon content is crucial for a lot of industries. In this module you have seen a variety of ways to measure Total Carbon TC, as well as the source of that carbon, whether it be organic in nature (TOC), or inorganic (TIC). This information is extremely important for several industries: from oil exploration, where information on carbon content is needed to evaluate a formation's production viability, to regulatory agencies, where carbon content and its origin are needed to ensure quality control and public safety.

TOC, TC, TIC measurements do have significant limitations. Mostly all techniques are destructive in nature, meaning that sample cannot be recovered. Further limitations include assumptions that have to be made in the measurement. In TOC measurement for example, assumptions that all TIC has been removed in pretreatments with acid have to be made, as well as that all organic carbon is completely oxidized to CO_2. In TIC measurements, it is assumed that all carbon sources are removed from the sample and detected. Several things can be done to promote these conditions so as to make such assumptions valid.

All measurements cost money, because TOC, TIC, and TC are all related, more frequently than not only two measurements are done, and the third value is found by using their relation to one another.

Bibliography

B. B. Bernard, H. Bernard, and J.M. Brooks: *Determination of Total Carbon, Total Organic Carbon and Inorganic Carbon in Sediments*, College Station, Texas, USA, DI-Brooks International and B&B Laboratories, Inc.

S. W. Kubala, D. C. Tilotta, M. A. Busch, and K. W. Busch, Determination of total inorganic carbon in aqueous samples with a flame infrared emission detector. *Anal. Chem.*, 1989, **61**, 1841.

S. J. Maguire-Boyle and A. R. Barron, Organic compounds in produced waters from shale gas wells. *Environ. Sci.: Processes Impacts*, 2014, **16**, 2237.

B. A. Schumacher, *Methods for the determination of Total Organic Carbon (TOC) in Soils and Sediments*. U.S. Environmental Protection Agency, Washington, DC, EPA/600/R-02/069 (NTIS PB2003-100822) (2002).

Z. A, Wang, S. N. Chu, and K. A. Hoering, High-frequency spectrophotometric measurements of total dissolved inorganic carbon in seawater. *Environ. Sci. Technol.*, 2013, **47**, 7840.

Chapter 11: Atomic Fluorescence Spectroscopy

Danielle Michaud, Lei Li and Andrew R. Barron

Introduction

Atomic fluorescence spectroscopy (AFS) is a method that was invented by Winefordner (Figure 11.1) and Vickers in 1964 as a means to analyze the chemical concentration of a sample. The idea is to excite a sample vapor with the appropriate UV radiation, and by measuring the emitting radiation, the amount of the specific element being measured could be quantified. In its most basic form, AFS consists of a UV light source to excite the sample, a monochromator, a detector and a readout device. Cold vapor atomic fluorescence spectroscopy (CVAFS) uses the same technique as AFS, but the preparation of the sample is adapted specifically to quantify the presence of heavy metals that are volatile, such as mercury, and allows for these elements to be measured at room temperature.

Figure 11.1: American chemist James D. Winefordner (1931-).

Fluorescence

Fluorescence is a process involving the emission of light from any substance in the excited states. Generally speaking, fluorescence is the emission of electromagnetic radiation (light) by the substance absorbed the different wavelength radiation. Its absorption and emission are illustrated in the

Jablonski diagram (Figure 11.2), a fluorophore is excited to higher electronic and vibrational state from ground state after excitation. The excited molecules can relax to lower vibrational state due to the vibrational relaxation and, then further retune to the ground state in the form of fluorescence emission.

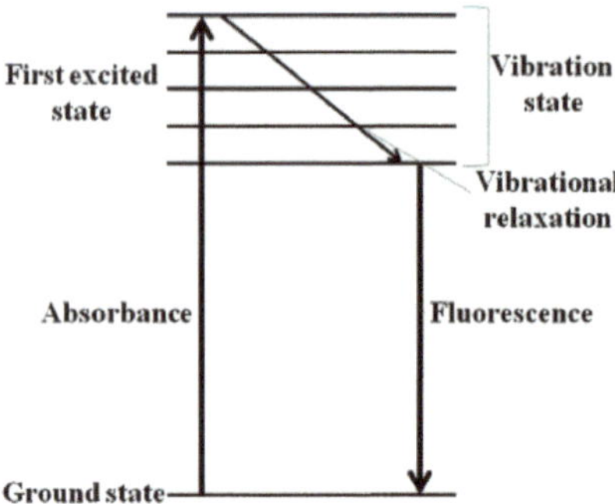

Figure 11.2: Jablonski diagram of fluorescence.

Instrumentation

Most spectrofluorometers can record both excitation and emission spectra. They mainly consist of four parts: light sources, monochromators, optical filters and detector (Figure 11.3).

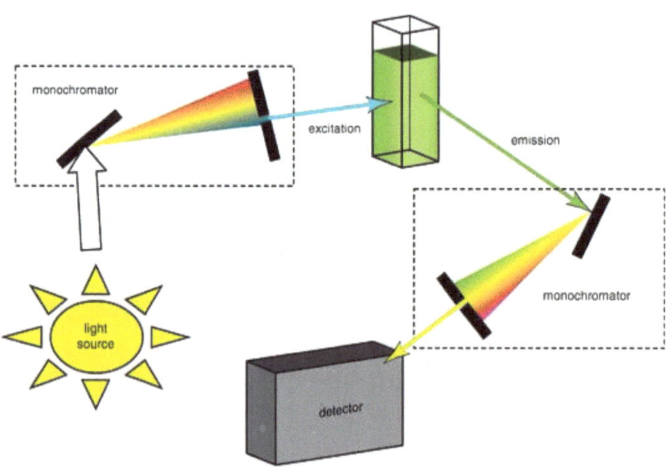

Figure 11.3: Schematic representation of a fluorescence spectrometer.

Light sources

Light sources that can emit wavelength of light over the ultraviolet and the visible range can provide the excitation energy. There are different light sources, including arc and incandescent xenon lamps, high-pressure mercury (Hg) lamps, Xe-Hg arc lamps, low pressure Hg and Hg-Ar lamps, pulsed xenon lamps, quartz-tungsten halogen (QTH) lamps, LED light sources, etc. The proper light source is chosen based on the application.

Monochromators

Prisms and diffraction gratings are two mainly used types of monochromators, which help to get the experimentally needed chromatic light with a wavelength range of 10 nm. Typically, the monochromators are evaluated based on dispersion, efficiency, and stray light level and resolution.

Optical filters

Optical filters are used in addition to monochromators in order to further purifying the light. There are two kinds of optical filters. The first one is the colored filter, which is the most traditional filter and is also divided into two categories: monochromatic filter and long-pass filter. The other one is thin film filter that is the supplement for the former one in the application and being gradually instead of colored filter.

Detector

An InGaAs array is the standard detector used in many spectrofluorometers. It can provide rapid and robust spectral characterization in the near-IR.

Cold vapor atomic fluorescence spectroscopy (CVAFS)

Cold vapor atomic fluorescence spectroscopy (CVAFS) uses the same technique as AFS, but the preparation of the sample is adapted specifically to quantify the presence of heavy metals that are volatile, such as mercury, and allows for these elements to be measured at room temperature (Figure 11.4)

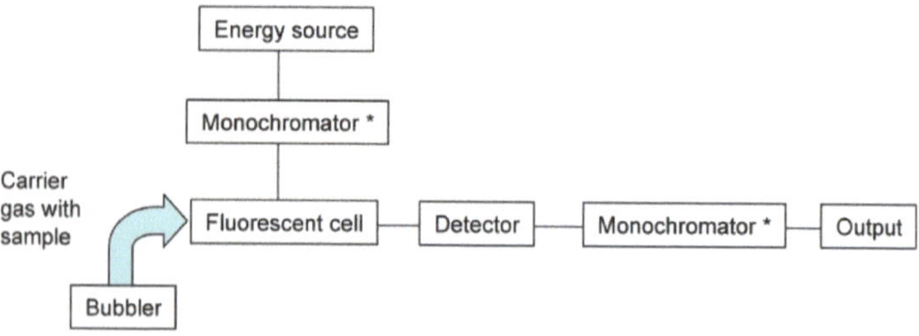

Figure 11.4: The basic setup for CVAFS. *The monochromator can be in either position in the scheme.

Theory

The theory behind CVAFS is that as the sample absorbs photons from the radiation source, it will enter an excited state. As the atom falls back into the ground state from its excited vibrational state(s), it will emit a photon, which can then be measured to determine the concentration. In its most basic sense, this process is represented by

$$P_F = \varphi P_{abs}$$

where P_F is the power given of as photons from the sample, P_{abs} is the power of the radiation absorbed by the sample, and φ is the proportionality factor of the energy lost due to collisions and interactions between the atoms present, and not due to photon emission.

Sample preparation

As is common with all forms of atomic fluorescence spectroscopy (AFS) and atomic absorption spectrometry (AES), the sample must be digested, usually with an acid, to break down the compounds so that all metal atoms in the sample are accessible to be vaporized. The sample is put into a bubbler (Figure 11.4), usually with an agent that will convert the element to its gaseous species. An inert gas carrier such as argon is then passed through the bubbler to carry the metal vapors to the fluorescence cell. It is important that the gas carrier is inert, so that the signal will only be absorbed and emitted by the sample in question and not the carrier gas.

Once the sample is loaded into the cell, a collimated (almost parallel) UV light source passes through the sample so that it will fluoresce. A monochromator is often used, either between the light source and the sample, or between the sample and the detector. These two different setups are referred to as excitation or emission spectrum, respectively. In an excitation spectrum, the light source is kept at a constant wavelength via the monochromator, and multiple wavelengths of emitted light are gathered, whereas in the emission spectrum, only the specified wavelength of light emitted from the sample is measured, but the sample is exposed to multiple wavelengths of light from the excitatory source. The fluorescence will be detected by a photomultiplier tube, which is extremely light sensitive, and a photodiode is used to convert the light into voltage or current, which can then in turn be interpreted into the amount of the chemical present.

Fluorescence molecular sensors

Fluorescence molecular sensor, one type of fluorescence molecular probe, can be fast, reversible response in the recognition process. There are four factors, selectivity, sensitivity, in-situ detection, and real time, that are generally used to evaluate the performance of the sensor. In this paper, four fundamental principles for design fluorescence molecular sensors are introduced.

Photoinduced electron transfer (PET)

Photoinduced electron transfer is the most popular principle in the design of fluorescence molecular sensors. The characteristic structure of PET sensors includes three parts as shown in Figure 11.5:

- The fluorophore absorbs the light and emits fluorescence signal.
- The receptor selectively interacts with the guest.
- A spacer connects the fluorophore and receptor together to form an integral system and successfully, effectively transfers the recognition information from receptor to fluorophore.

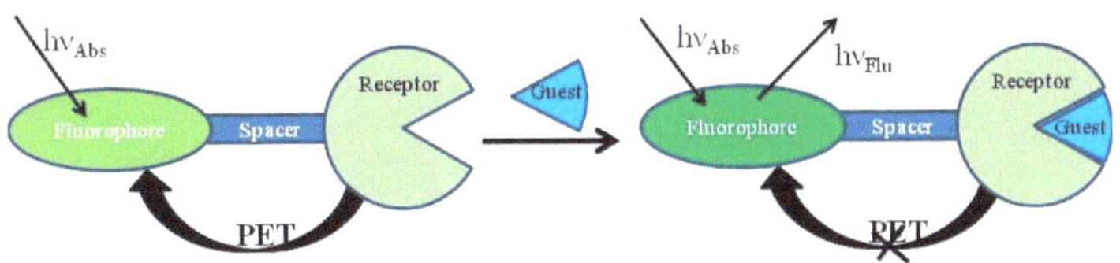

Figure 11.5: The general view of the principle of PET fluorescence molecular sensor.

In the PET sensors, photoinduced electron transfer makes the transfer of recognition information to fluorescence signal between receptor and fluorophore come true. Figure 1.91 shows the detailed process of how PET works in the fluorescence molecular sensor. The receptor could provide the electron to the vacated electoral orbital of the excited fluorophore. The excited electron in the fluorophore could not come back the original orbital, resulting in the quenching of fluorescence emission. The coordination of receptor and guest decreased the electron donor ability of receptor reduced or even disrupted the PET process, then leading to the enhancement of intensity of fluorescence emission. Therefore, the sensors had weak or no fluorescence emission before the coordination. However, the intensity of fluorescence emission would increase rapidly after the coordination of receptor and gust.

Intramolecular charge transfer (ICT)

Intramolecular charge transfer (ICT) is also named photoinduced charge transfer. The characteristic structure of ICT sensors includes only the fluorophore and recognition group, but no spacer. The recognition group directly binds to the fluorophore. The electron withdrawing or electron donating substituents on the recognition group plays an important role in the recognition. When the recognition happens, the coordination between the recognition group and guest affects the electron density in the fluorophore, resulting in the change of fluorescence emission in the form of blue shift or red shift.

Excimer

When the two fluorophores are in the proper distance, an intermolecular excimer can be formed between the excited state and ground state. The fluorescence emission of the excimer is different with the monomer and mainly in the form of new, broad, strong, and long wavelength emission without fine structures. The proper distance determines the formation of excimer, therefore modulation of the distance between the two fluorophores becomes crucial in the design of the sensors based on this mechanism. The fluorophores have long lifetime in the singlet state to be easily forming the excimers. They are often used in such sensors.

Fluorescence resonance energy transfer (FRET)

FRET is a popular principle in the design of the fluorescence molecular sensor. In one system, there are two different fluorophores, in which one acts as a donor of excited state energy to the receptor of the other. As shown in Figure

11.6, the receptor accepts the energy from the excited state of the donor and gives the fluorescence emission, while the donor will return back to the electronic ground state. There are three factors affecting the performance of FRET. They are the distance between the donor and the acceptor, the proper orientation between the donor emission dipole moment and acceptor absorption moment, and the extent of spectral overlap between the donor emission and acceptor absorption spectrum (Figure 11.7).

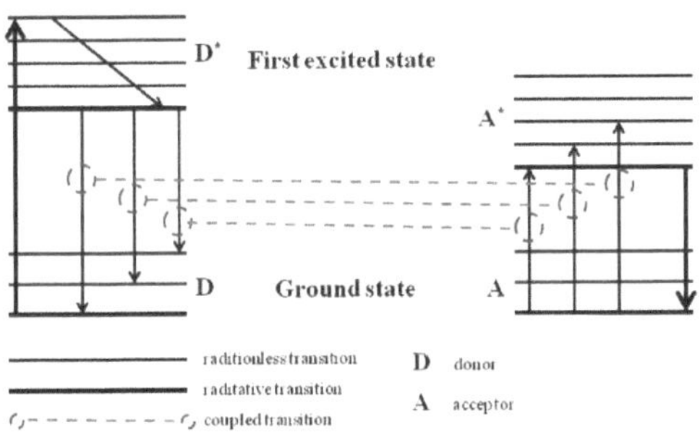

Figure 11.6: A schematic fluorescence resonance energy transfer system.

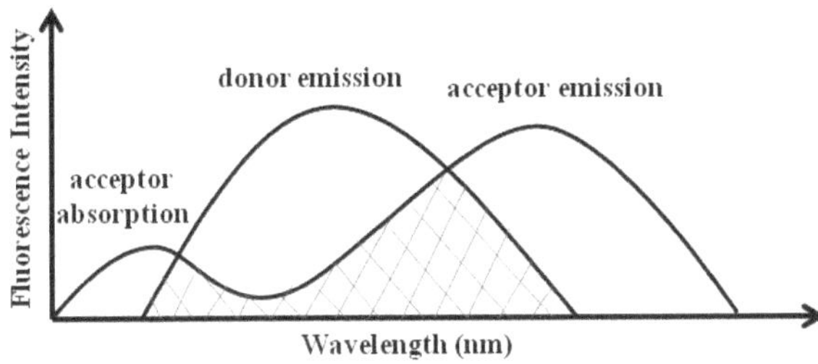

Figure 11.7: Diagram showing the spectral overlap for fluorescence resonance energy transfer system.

Detection of mercury: a case study

Mercury pollution has become a global problem and seriously endangers human health. Inorganic mercury can be easily released into the environment through a variety of anthropogenic sources, such as the coal mining, solid waste incineration, fossil fuel combustion, and chemical manufacturing. It can also be released through the nonanthropogenic sources in the form of forest fires, volcanic emissions, and oceanic emission.

Mercury can be easily transported into the atmosphere as the form of the mercury vapor. The atmospheric deposition of mercury ions leads to the accumulation on plants, in topsoil, in water, and in underwater sediments. Some prokaryotes living in the sediments can convert the inorganic mercury into methylmercury, which can enter food chain and finally is ingested by human.

Mercury poisoning can damage the nervous system, kidneys, and also fetal development in pregnant women, so it is important to evaluate the levels of mercury present in our environment. Some of the more common sources of mercury are in the air (from industrial manufacturing, mining, and burning coal), the soil (deposits, waste), water (byproduct of bacteria, waste), and in food (especially seafood). Although regulation for food, water and air mercury content differs, EPA regulation for mercury content in water is the lowest, and it cannot exceed 2 ppb (27 μg/L).

Cold vapor atomic fluorescence spectroscopy.

Sample preparation

In 1972, J. F. Kopp et al. first published a method to detect minute concentrations of mercury in soil, water, and air using gold amalgamation and cold vapor atomic fluorescence spectroscopy. While atomic absorption can also measure mercury concentrations, it is not as sensitive or selective as cold vapor atomic fluorescence spectroscopy (CVAFS).

The sample is put in the bubbler with a reducing agent such as stannous chloride ($SnCl_2$) so that Hg^0 is the only state present in the sample. Once the mercury is in its elemental form, the argon enters the bubbler through a gold trap, and carries the mercury vapors out of the bubbler to the first gold trap, after first passing through a soda lime (mixture of $Ca(OH)_2$, $NaOH$, and KOH) trap where any remaining acid or water vapors are caught. After all the mercury from the sample is absorbed by the first gold trap, it is heated to 450 °C, which causes the mercury absorbed onto the gold trap to be carried by the argon gas to the second gold trap. Once the mercury from the sample has been absorbed by the second trap, it is heated to 450 °C, releasing the mercury to be carried by the argon gas into the fluorescence cell, where light at a wavelength of 253.7 nm will be used for mercury samples. The detection limit for mercury using gold amalgamation and CVAFS is around 0.05 ng/L, but the detection limit will vary due to the equipment being used, as well as human error.

Calculating CVAFS concentrations

A standard solution of mercury should be made, and from this solution will be used to make at least five different standard solutions by dilution. Depending on the detection limit and what is being analyzed, the concentrations in the standard solutions will vary. Note that what other chemicals the standard solutions contain will depend upon how the sample is digested.

For example, a 1.00 g/mL Hg (1 ppm) working solution is made, and by dilution, five standards are made from the working solution, at 5.0, 10.0, 25.0, 50.0, and 100.0 ng/L (ppt). If these five standards give peak heights of 10 units, 23 units, 52 units, 110 units, and 207 units, respectively, then

$$CF_x = A_x/C_x$$

is used to calculate the calibration factor, where CF_x is the calibration factor, A_x is the area of the peak or peak height, and C_x is the concentration in ng/L of the standard,

$$10/5.0 \text{ ng/L} = 2.00 \text{ units.L/ng}$$

The calibration factors for the other four standards are calculated in the same fashion: 2.30, 2.08, 2.20, and 2.07, respectively. The average of the five calibration factors is then taken,

$$CF_m = (2.00 + 2.30 + 2.08 + 2.20 + 2.07)/5 = 2.13 \text{ units.L/ng}$$

Now to calculate the concentration of mercury in the sample,

$$[Hg] \text{ (ng/L)} = (A_s/CF_m).(V_{std}/V_{smp})$$

where, A_s is the area of the peak of the sample, CF_m is the mean calibration factor, V_{std} is the volume of the standard solution minus the reagents added, and V_{smp} is the volume of the initial sample (total volume minus volume of reagents added). If A_s is measured at 49 units, $V_{std} = 0.47$ L, and $V_{smp} = 0.26$ L, then the concentration can be calculated,

(49 units/2.13 units.L/ng).(0.47 L/0.26 L) = 43.2 ng/L of Hg present

Sources of error

Contamination from the sample collection is one of the biggest sources of error: if the sample is not properly collected or hands/gloves are not clean, this can tamper with the concentration. Also, making sure the glassware and equipment is clean from any sources of contamination. Furthermore, sample vials that are used to store mercury-containing samples should be made out of borosilicate glass or fluoropolymer, because mercury can leach or absorb other materials, which could cause an inaccurate concentration reading.

Fluorescence molecular sensors

PET fluorescence sensor

As a PET sensor 2-{5-[(2-{[*bis*-(2-ethylsulfanyl-ethyl)-amino]-methyl}-phenylamino)-methyl]-2-chloro-6-hydroxy-3-oxo-3H-xanthen-9-yl}-benzoic acid (MS1) (Figure 11.8) shows good selectivity for mercury ions in buffer solution (pH = 7, 50 mM PIPES, 100 mM KCl). Upon the increase of the concentration of Hg^{2+} ions, the coordination between the sensor and Hg^{2+} ions disrupted the PET process, leading to the increase of the intensity of fluorescence emission with slight red shift to 528 nm. Sensor MS1 also showed good selectivity for Hg^{2+} ions over other cations of interest; moreover, it had good resistance to the interference from other cations when detected Hg^{2+} ions in the mixture solution excluding Cu^{2+} ions.

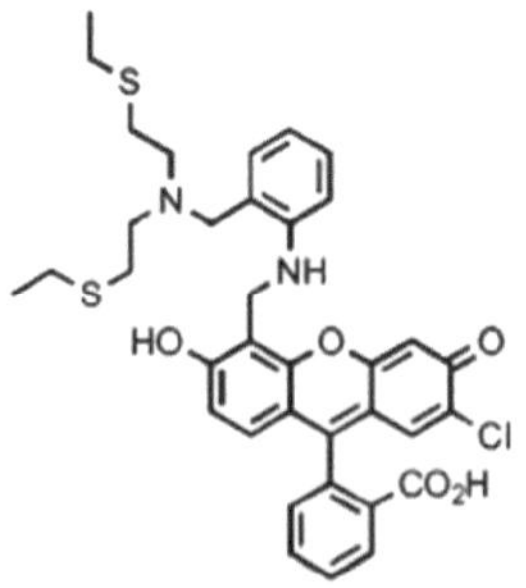

Figure 11.8: Structure of the PET fluorescence sensor 2-{5-[(2-{[bis-(2-ethyl-sulfanyl-ethyl)-amino]- methyl}-phenylamino)-methyl]-2-chloro-6-hydroxy-3-oxo-3H-xanthen-9-yl}-benzoic acid.

ICT fluorescence sensor

2,2',2,2'-(3-(benzo[d]thiazol-2-yl)-2-oxo-2-H-chromene-6,7-diyl)*bis*-(azane-triyl)*tetrakis*(N-(2- hydroxyethyl)acetamide) (RMS) (Figure 11.9) has been shown to be an ICT fluorescence sensor. with the gradual increase of the concentration of Hg^{2+} ions, fluorescence emission spectra revealed a significant blue shift, which was about 100-nm emission band shift from 567 to 475 nm in the presence of 40 equivalent of Hg^{2+} ions. The fluorescence change came from the coexistence of two electron-rich aniline nitrogen atoms in the electron-donating receptor moiety, which prevented Hg^{2+} ions ejection from them simultaneously in the excited ICT fluorophore. Sensor RMS also showed good selectivity over other cations of interest. As shown in Figure 11.10, it is easy to find that only Hg^{2+} ions can modulate the fluorescence of RMS in a neutral buffered water solution.

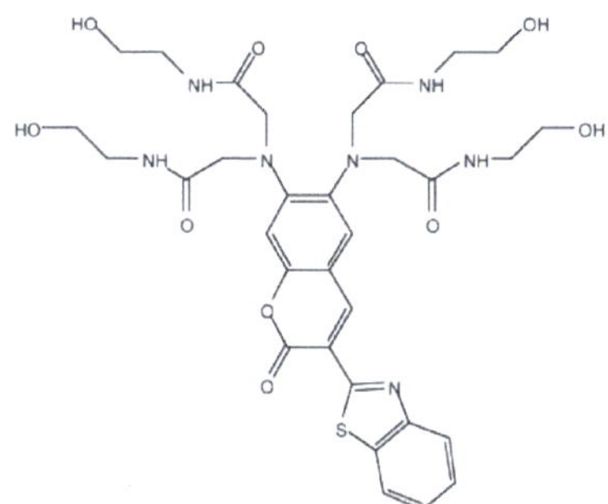

Figure 11.9: Structure of the ICT fluorescence sensor 2,2',2,2'-(3-(benzo[d]thia-zol-2-yl)-2-oxo-2-H-chromene-6,7-diyl)*bis*(azanetriyl)*tetrakis*(N-(2-hydroxyethyl)acetamide) (RMS).

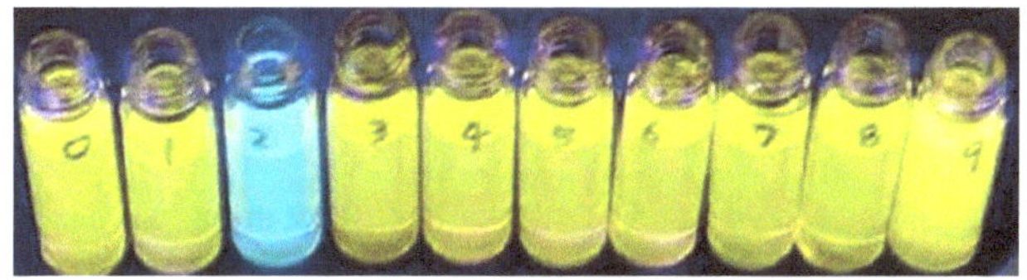

Figure 11.10: Fluorescence response of 10 μM RMS in the presence of 20 equivalent of different cations of interest at the same condition: (0) control, (1) Cd^{2+}, (2) Hg^{2+}, (3) Fe^{3+}, (4) Zn^{2+}, (5) Ag^+, (6) Co^{2+}, (7) Cu^{2+}, (8) Ni^{2+}, and (9) Pb^{2+}. Adapted from J. B. Wang, X. H. Qian, and J. G. Cui, Detecting Hg^{2+} ions with an ICT fluorescent sensor molecule: remarkable emission spectra shift and unique selectivity. *J. Org. Chem.*, 2006, 71, 4308. Copyright: American Chemical Society (2006).

Excimer fluorescence sensors

The (NE,N'E)-2,2'-(ethane-1,2-diyl-*bis*(oxy))*bis*(N-(pyren-4-ylmethylene)-aniline) (BA) (Figure 11.11) is the excimer fluorescence sensor. When BA existed without mercury ions in the mixture of HEPES-CH$_3$CN (80:20, v/v, pH 7.2), it only had the weak monomer fluorescence emission. Upon the increase of the concentration of mercury ions in the solution of BA, a strong excimer fluorescence emission at 462 nm appeared and increased with the change of the concentration of mercury ions. BA showed good selectivity for mercury ions. Moreover, it had good resistance to the interference when detecting mercury ions in the mixture solution.

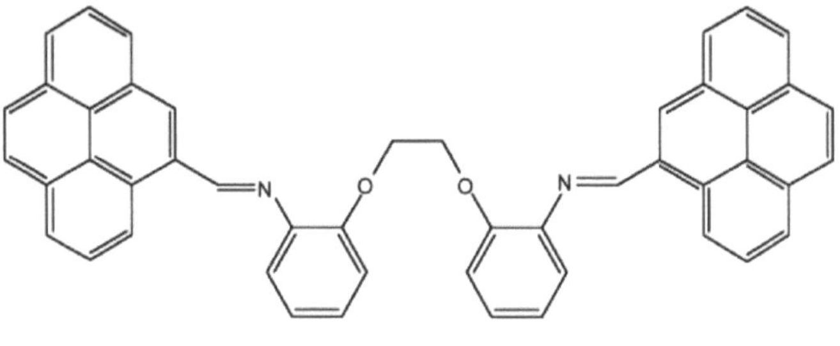

Figure 11.11: Structure of the excimer fluorescence sensor (NE,N'E)-2,2'-(ethane-1,2-diyl-*bis*(oxy))*bis*(N-(pyren-4-ylmethylene) aniline) (BA).

FRET fluorescence sensors

The calix[4]arene derivative bearing two pyrene and rhodamine fluorophores (CPR) (Figure 11.12) is a characteristic FRET fluorescence sensor. Fluorescence titration experiment of CPR (10.0 µM) with Hg^{2+} ions was carried out in CHCl$_3$/CH$_3$CN (50:50, v/v) with an excitation of 343 nm. Upon gradual increase the concentration of Hg^{2+} ions in the solution of CPR, the increased fluorescence emission of the ring-opened rhodamine at 576 nm was observed with a concomitantly declining excimer emission of pyrene at 470 nm. Moreover, an isosbestic point centered at 550 nm appeared. This change in the fluorescence emission demonstrated that an energy from the pyrene excimer transferred to rhodamine, resulting from the trigger of Hg^{2+} ions. CPR had good resistance to other cations of interest when detected Hg^{2+} ions, though Pb^{2+} ions had little interference in this process.

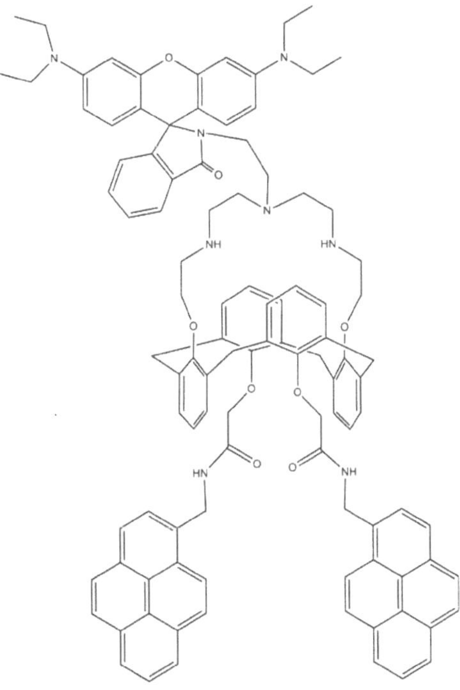

Figure 11.17: Structure of the FRET fluorescence sensor calix[4]arene derivative (CPR) bearing two pyrene and rhodamine fluorophores.

Bibliography

J. S. Kim and D. T. Quang, Calixarene-derived fluorescent probes. *Chem. Rev.*, 2007, **107**, 3780.

J. F. Kopp, M. C. Longbottom, and L. B. Lobring, "Cold vapor" method for determining mercury. *AWWA*, 1972, **64**, 20.

J. R. Lakowicz, *Principles of Fluorescence Spectroscopy*, 3rd edn., Springer (2006).

Y. H. Lee, M. H. Lee, J. F. Zhang, and J. S. Kim, Pyrene excimer-based calix[4]arene FRET chemosensor for mercury(II). *J. Org. Chem.*, 2010, **75**, 7159.

Method 245.7: *Mercury in Water by Cold Vapor Atomic Fluorescence Spectrometry*, Revision 2.0. EPA. 2005.

E. M. Nolan and S. J. Lippard, A "turn-on" fluorescent sensor for the selective detection of mercuric ion in aqueous media. *J. Am. Chem. Soc.*, 2003, **125**, 14270.

E. M. Nolan and S. J. Lippard, Tools and tactics for the optical detection of mercuric ion. *Chem. Rev.*, 2008, **108**, 3443.

J. B. Wang, X. H. Qian, and J. G. Cui, Detecting Hg^{2+} ions with an ICT fluorescent sensor molecule: remarkable emission spectra shift and unique selectivity. *J. Org. Chem.*, 2006, **71**, 4308.

J. D. Winefordner and T. J. Vickers. Atomic fluorescence spectroscopy as a means of chemical analysis. *Anal. Chem.*, 1964, **36**, 161.

Y. Zhou, C. Y. Zhu, X. S. Gao, X. Y. You, and C. Yao, Hg^{2+}-selective ratiometric and "off−on" chemosensor based on the azadiene−pyrene derivative. *Org. Lett.*, 2010, **12**, 2566.

Chapter 12: Energy Dispersive X-ray Spectroscopy

Graham Piburn and Andrew R. Barron

Introduction

Energy-dispersive X-ray spectroscopy (EDX or EDS) is an analytical technique used to probe the composition of a solid materials. Several variants exist, but the all rely on exciting electrons near the nucleus, causing more distant electrons to drop energy levels to fill the resulting holes. Each element emits a different set of X-ray frequencies as their vacated lower energy states are refilled, so measuring these emissions can provide both qualitative and quantitative information about the near-surface makeup of the sample. However, accurate interpretation of this data is dependent on the presence of high-quality standards, and technical limitations can compromise the resolution.

Physical underpinnings

In the quantum mechanical model of the atom, an electron's energy state is defined by a set of quantum numbers. The primary quantum number, n, provides the coarsest description of the electron's energy level, and all the sublevels that share the same primary quantum number are sometimes said to comprise an energy shell. Instead of describing the lowest-energy shell as the $n = 1$ shell, it is more common in spectroscopy to use alphabetical labels: The K shell has $n = 1$, the L shell has $n = 2$, the M shell has $n = 3$, and so on. Subsequent quantum numbers divide the shells into subshells: one for K, three for L, and five for M. Increasing primary quantum numbers correspond with increasing average distance from the nucleus and increasing energy (Figure 12.1). An atom's core shells are those with lower primary quantum numbers than the highest occupied shell, or valence shell.

Transitions between energy levels follow the law of conservation of energy. Excitation of an electron to a higher energy state requires an input of energy from the surroundings, and relaxation to a lower energy state releases energy to the surroundings. One of the most common and useful ways energy can be transferred into and out of an atom is by electromagnetic radiation. Core shell transitions correspond to radiation in the X-ray portion of the spectrum; however, because the core shells are normally full by definition, these transitions are not usually observed.

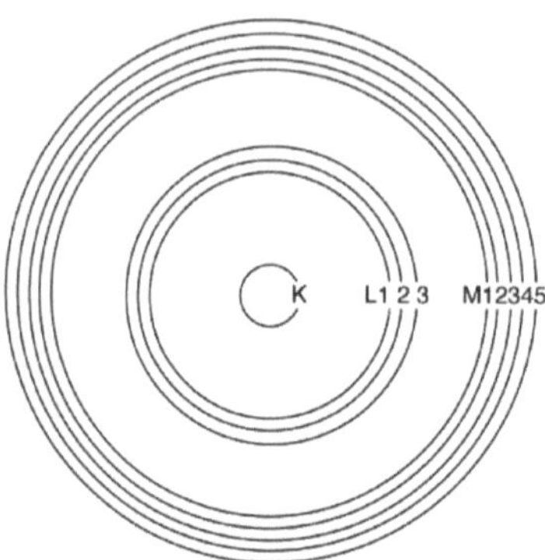

Figure 12.1: A diagram of the core electronic energy levels of an atom, with the lowest energy shell, K, nearest the nucleus. Circles are used here for convenience they are not meant to represent the shapes of the electron's orbitals.

X-ray spectroscopy uses a beam of electrons or high-energy radiation (see instrument variations, below) to excite core electrons to high energy states, creating a low-energy vacancy in the atoms' electronic structures. This leads to a cascade of electrons from higher energy levels until the atom regains a minimum-energy state. Due to conservation of energy, the electrons emit X-rays as they transition to lower energy states. It is these X-rays that are being measured in X-ray spectroscopy.

The energy transitions are named using the letter of the shell where ionization first occurred, a Greek letter denoting the group of lines that transition belongs to, in order of decreasing importance, and a numeric subscript ranking the peak's the intensity within that group. Thus, the most intense peak resulting from ionization in the K shell would be $K\alpha_1$ (Figure 12.2). Since each element has a different nuclear charge, the energies of the core shells and, more importantly, the spacing between them vary from one element to the next. While not every peak in an element's spectrum is exclusive to that element, there are enough characteristic peaks to be able to determine composition of the sample, given sufficient resolving power.

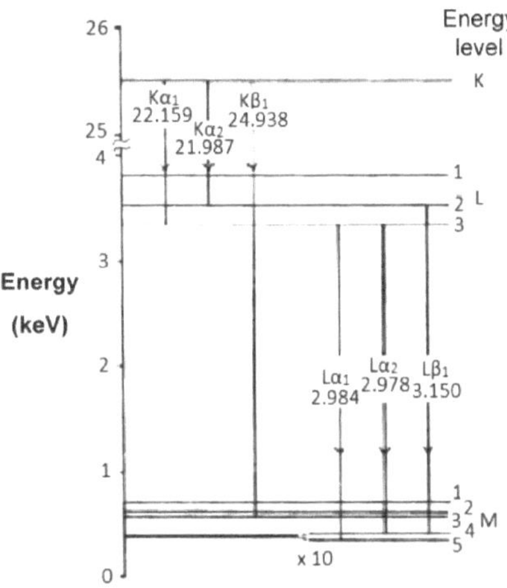

Figure 12.2: A diagram of the energy transitions after the excitation of a gold atom. The arrows show the direction the vacancy moves when the higher energy electrons move down to refill the core.

Instrumentation and sample preparation

Instrument variations

There are two common methods for exciting the core electrons of the surface atoms. The first is to use a high-energy electron beam like the one in a scanning electron microscope (SEM). The beam is produced by an electron gun, in which electrons emitted thermionically from a hot cathode are guided down the column by an electric field and focused by a series of negatively charged lenses. X-rays emitted by the sample strike a lithium-drifted silicon p-i-n junction plate. This promotes electrons in the plate into the conduction band, inducing a voltage proportional to the energy of the impacting X-ray which generally falls between about 1 and 10 keV. The detector is cooled to liquid nitrogen temperatures to reduce electronic noise from thermal excitations.

It is also possible to use X-rays to excite the core electrons to the point of ionization. In this variation, known as energy-dispersive X-ray fluorescence analysis (EDXRFA or XRF), the electron column is replaced by an X-ray tube and the X-rays emitted by the sample in response to the bombardment are called secondary X-rays, but these variants are otherwise identical.

Regardless of the excitation method, subsequent interactions between the emitted X-rays and the sample can lead to poor resolution in the X-ray spectrum, producing a Gaussian-like curve instead of a sharp peak. Indeed, this spreading of energy within the sample combined with the penetration of the electron or X-ray beam leads to the analysis of a roughly 1 μm^3 volume instead of only the surface features. Peak broadening can lead to overlapping peaks and a generally misleading spectrum. In cases where a normal EDS spectrum is inadequately resolved, a technique called wavelength-dispersive X-ray spectroscopy (WDS) can be used. The required instrument is very similar to the ones discussed above and can use either excitation method. The major difference is that instead of having the X-rays emitted by the sample hit the detector directly, they first encounter an analytical crystal of know lattice dimensions. Bragg's law predicts that the strongest reflections of the crystal will occur for wavelengths such that the path difference between a ray reflecting from consecutive layers in the lattice is equal to an integral number of wavelengths. This is represented mathematically as

$$n\lambda = 2d\ \sin\theta$$

where n is an integer, λ is the wavelength of impinging light, d is the distance between layers in the lattice, and θ is the angle of incidence. The relevant variables for the equation are labeled in Figure 13.3.

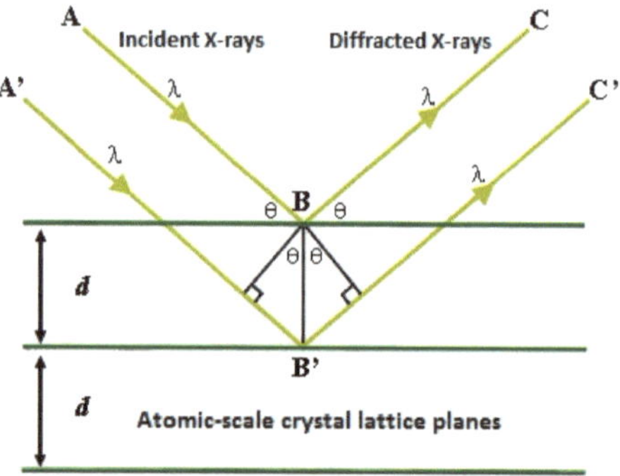

Figure 12.3: A diagram of a light beam impinging on a crystal lattice. If the light meets the criterion $n\lambda = 2d\ \sin(\theta)$, Bragg's law predicts that the waves reflecting of each layer of the lattice interfere constructively, leading to a strong signal. Reproduced from D. Henry, N. Eby, J. Goodge, and D. Mogk, *X-ray reflection in accordance with Bragg's Law,* http://serc.carleton.edu/research_education/geochemsheets/BraggsLaw.html**.**

By moving the crystal and the detector around the Rowland circle, the spectrometer can be tuned to examine specific wavelengths (Figure 12.4). Generally, an initial scan across all wavelengths is taken first, and then the instrument is programmed to more closely examine the wavelengths that produced strong peaks. The resolution available with WDS is about an order of magnitude better than with EDS because the analytical crystal helps filter out the noise of subsequent, non-characteristic interactions. For clarity, X-ray spectroscopy will be used to refer to all of the technical variants just discussed, and points made about EDS will hold true for XRF unless otherwise noted.

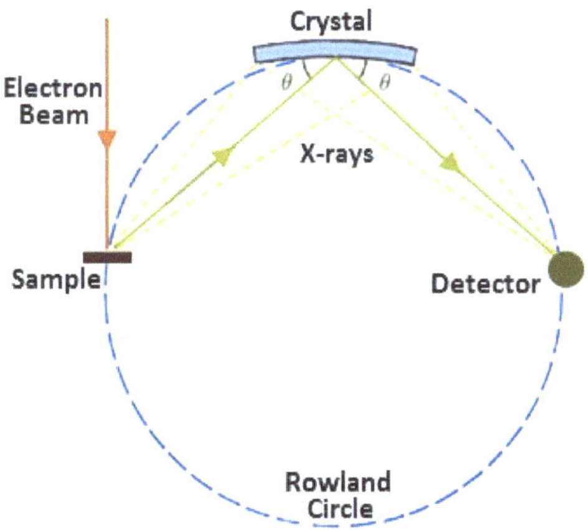

Figure 12.4: A schematic of a typical WDS instrument. The analytical crystal and the detector can be moved around an arc known as the Rowland Circle. This grants the operator the ability to change the angle between the sample, the crystal, and the detector, thereby changing the X-ray wavelength that would satisfy Bragg's law. The sample holder is typically stationary. Reproduced from D. Henry and J. Goodge, Wavelength-dispersive X-ray spectroscopy (WDS), http://serc.carleton.edu/research_education/geochemsheets/wds.html.

Sample preparation

Compared with some analytical techniques, the sample preparation required for X-ray spectroscopy or any of the related methods just discussed is trivial. The sample must be stable under vacuum, since the sample chamber is evacuated to prevent the atmosphere from interfering with the electron beam or X-rays. It is also advisable to have the surface as clean as possible; X-ray spectroscopy is a near-surface technique, so it should analyze the desired material for the most part regardless, but any grime on the surface will throw of the composition calculations. Simple qualitative readings can be obtained from a solid of any thickness, as long as it fits in the machine, but for reliable quantitative measurements, the sample should be shaved as thin as possible.

Data interpretation

Qualitative analysis, the determination of which elements are present in the sample but not necessarily the stoichiometry, relies on empirical standards. The energies of the commonly used core shell transitions have been tabulated for all the natural elements. Since combinations of elements can act differently than a single element alone, standards with compositions as similar as possible to the suspected makeup of the sample are also employed. To determine the sample's composition, the peaks in the spectrum are matched with peaks from the literature or standards.

Quantitative analysis, the determination of the sample's stoichiometry, needs high resolution to be good enough that the ratio of the number of counts at each characteristic frequency gives the ratio of those elements in the sample. It takes about 40,000 counts for the spectrum to attain a 2σ precision of $\pm 1\%$. It is important to note, however, that this is not necessarily the same as the empirical formula, since not all elements are visible. Spectrometers with a beryllium window between the sample and the detector typically cannot detect anything lighter than sodium. Spectrometers equipped with polymer-based windows can quantify elements heavier than beryllium. Either way, hydrogen cannot be observed by X-ray spectroscopy.

X-ray spectra are presented with energy in keV on the x-axis and the number of counts on the y-axis.

The EDX spectra of biotite (Figure 12.5) is shown in Figure 12.6. Biotite is a mineral similar to mica which has the approximate chemical formula $K(Mg,Fe)_3AlSi_3O_{10}(F,OH)_2$ (Figure 12.7). Strong peaks for manganese, aluminum, silicon, potassium, and iron can be seen in the spectrum. The lack of visible hydrogen is expected, and the absence of oxygen and fluorine peaks suggests the instrument had a beryllium window. The titanium peak is small and unexpected, so it may only be present in trace amounts.

Figure 12.5: Image of a biotite aggregate.

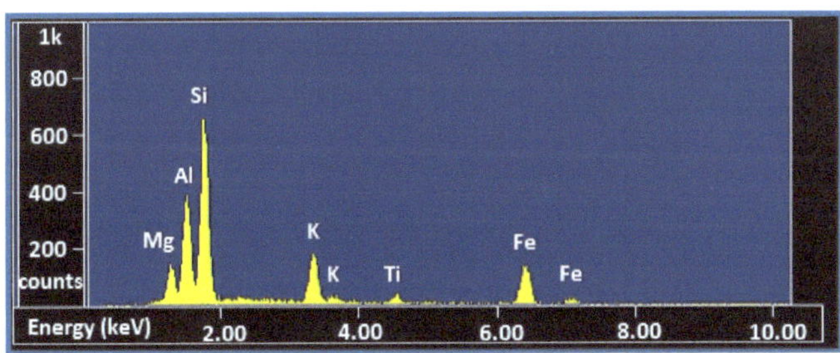

Figure 12.6: EDS spectrum of biotite in which silicon, aluminum, manganese, potassium, magnesium, iron, and titanium are all identifiable, though titanium appears to be only a trace component. Reproduced from J. Goodge, *Energy-dispersive X-ray spectroscopy (EDS)*, http://serc.carleton.edu/research_education/geochemsheets/eds.html.

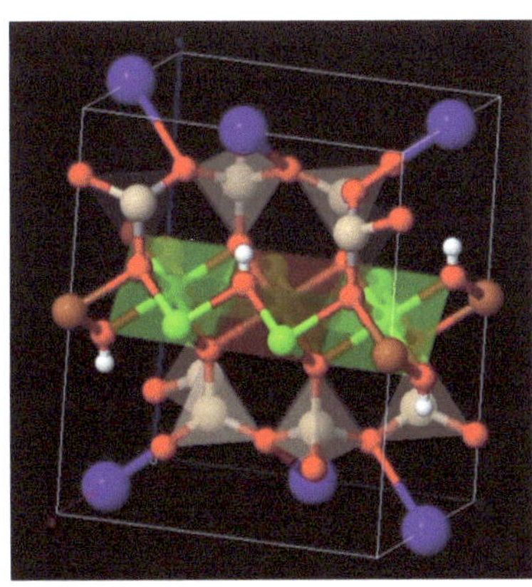

Figure 12.7: Ball and stick structure of biotite. Atom colors: light-grey = Si or Al; red = O; tan = Fe; green = Mg; purple = K; white = H.

K309 is a mix of glass developed by the National Institute for Standards and Technology (NIST) as a standard. The EDX spectrum (Figure 12.8) shows that it contains significant amounts of silicon, aluminum, calcium, oxygen, iron, and barium. The large peak at the far left is the carbon signal from the carbon substrate the glass was placed on. β

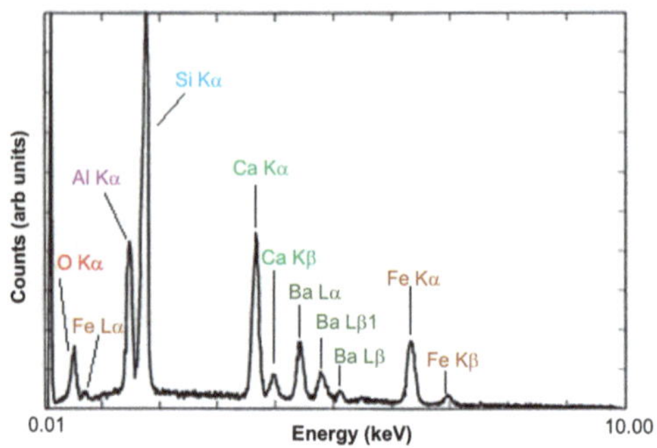

Figure 12.8: EDS spectrum of NIST K309 glass. Silicon, aluminum, barium, calcium, iron, and oxygen are identifiable in the spectrum. Adapted from J. Goldstein, D. Newbury, D. Joy, C. Lyman, P. Echlin, E. Lifshin, L. Sawyer, and J. Michael, *Scanning Electron Microscopy and X-ray Microanalysis*, 3rd edn., Springer, New York (2003).

Limitations

X-ray spectroscopy is incapable of seeing elements lighter than boron. This is a problem given the abundance of hydrogen in natural and man-made materials. The related techniques X-ray photoelectron spectroscopy (XPS) and Auger spectroscopy are able to detect Li and Be but are likewise unable to measure hydrogen.

X-ray spectroscopy relies heavily on standards for peak identification. Because a combination of elements can have noticeably different properties from the individual constituent elements in terms of X-ray fluorescence or absorption, it is important to use a standard as compositionally similar to the sample as possible. Naturally, this is more difficult to accomplish when examining new materials, and there is always a risk of the structure of the sample being appreciably different than expected.

The energy-dispersive variants of X-ray spectroscopy sometimes have a hard time distinguishing between emissions that are very near each other in energy or distinguishing peaks from trace elements from background noise. Fortunately, the wavelength-dispersive variants are much better at both of these. The rough, stepwise curve in Figure 12.9 represents the EDS spectrum of molybdenite (MoS_2), a mineral with a layered structure (Figure 12.10) Broadened peaks make it difficult to distinguish the molybdenum signals from the sulfur ones. Because WDS can select specific wavelengths, it has much better resolution and can pinpoint the separate peaks more accurately.

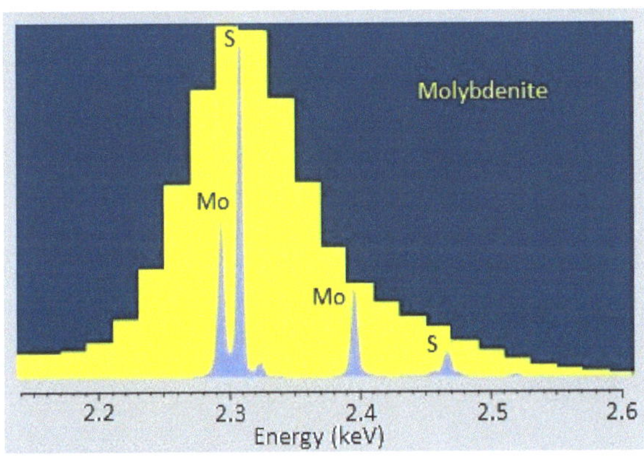

Figure 12.9: A comparison of the EDS (yellow) and WDS spectra (light blue) of a sample of molybdenite (MoS_2). The sulfur and molybdenum peaks are unresolved in the EDS spectrum, but are sharp and distinct in the WDS spectrum. Reproduced from Oxford Instruments, *The power of WDS sensitivity and resolution*, http://www.x-raymicroanalysis.com/x-ray-microanalysis-explained/pages/detectors/wave1.htm.

Figure 12.10: The structure of molybdenite (MoS_2) showing the layered structure. Atom colors: yellow = S; turquoise = Mo.

Bibliography

N. A. Baily and G. Kramer, The lithium-drifted silicon p-i-n junction as an X-ray and gamma-ray dosimeter. *Radiat. Res.*, 1964, **22**, 53.

J. Goldstein, D. Newbury, D. Joy, C. Lyman, P. Echlin, E. Lifshin, L. Sawyer, and J. Michael, *Scanning Electron Microscopy and X-ray Microanalysis*, 3rd edn., Springer, New York (2003).

J. Goodge, *Energy-Dispersive X-ray Spectroscopy (EDS)*, http://serc.carleton.edu/research_education/geochemsheets/.

K. F. J. Heinrich, *Electron Beam Microanalysis*, Van Nostrand Reinhold, New York (1981).

D. Henry and J. Goodge, *Wavelength-Dispersive X-ray Spectroscopy (WDS)*, http://serc.carleton.edu/research_education/geochemsheets/wds.html.

D. Henry, N. Eby, J. Goodge, and D. Mogk, *X-ray Reflection in Accordance with Bragg's Law*, http://serc.carleton.edu/research_education/geochemsheets/BraggsLaw.html.

Introduction to Energy Dispersive X-ray Spectroscopy http://micron.ucr.edu/public/manuals/EDS-intro.pdf.

Oxford Instruments, *The power of WDS sensitivity and resolution*, http://www.x-ray-microanalysis.com/x-ray-microanalysis-explained/pages/detectors/wave1.htm.

C. Whiston, *X-ray Methods*, Published on behalf of ACOL, Thames Polytechnic, London, by Wiley, New York (1987).

S. Zhang, L. Li, and A. Kumar, *Materials Characterization Techniques*, CRC, Boca Raton (2009).

Chapter 13: X-ray Photoelectron Spectroscopy

Lauren Harrison and Andrew R. Barron

Introduction

X-Ray photoelectron spectroscopy (XPS), also known as electron spectroscopy for chemical analysis (ESCA), is one of the most widely used surface techniques in materials science and chemistry. It allows the determination of atomic composition of the sample in a non-destructive manner, as well as other chemical information, such as binding constants, oxidation states and speciation. The sample under study is subjected to irradiation by a high energy X-ray source. The X-rays penetrate only 5 - 20 Å into the sample, allowing for surface specific, rather than bulk chemical, analysis. As an atom absorbs the X-rays, the energy of the X-ray will cause a K-shell electron to be ejected, as illustrated by Figure 13.1. The K-shell is the lowest energy shell of the atom. The ejected electron has a kinetic energy (KE) that is related to the energy of the incident beam ($h\nu$), the electron binding energy (BE), and the work function of the spectrometer (φ),

$$BE = h\nu - KE - \phi_s$$

Thus, the binding energy of the electron can be calculated.

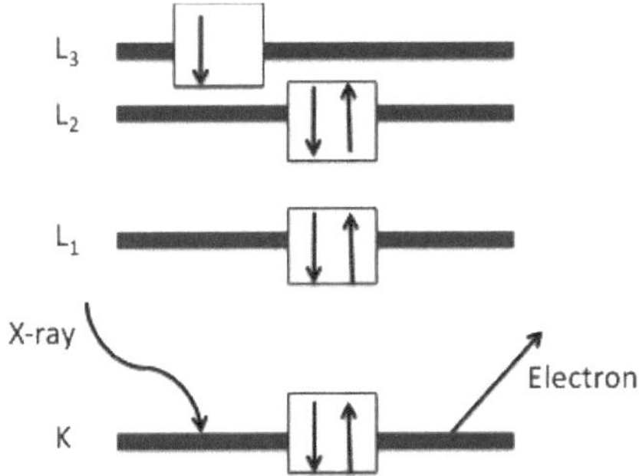

Figure 13.1: Excitation of an electron from an atom's K-shell.

Table 13.1 shows the binding energy of the ejected electron, and the orbital from which the electron is ejected, which is characteristic of each element. The number of electrons detected with a specific binding energy is proportional to the number of corresponding atoms in the sample. This then provides the percent of each atom in the sample.

Element	Peak	Binding energy (eV)
Carbon	C 1s	284.5 - 285.1
Nitrogen	N 1s	396.1 - 400.5
Oxygen	O 1s	526.2 - 533.5
Silicon	Si 2p	98.8 - 99.5
Sulfur	S $2p_{3/2}$	164.0 - 164.3
Iron	Fe $2p_{3/2}$	706.8 - 707.2
Gold	Au $4f_{7/2}$	83.8 - 84.2

Table 13.1: Binding energies for select elements in their elemental forms.

The chemical environment and oxidation state of the atom can be determined through the shifts of the peaks within the range expected (Table 13.2). If the electrons are shielded then it is easier, or requires less energy, to remove them from the atom, i.e., the binding energy is low. The corresponding peaks will shift to a lower energy in the expected range. If the core electrons are not shielded as much, such as the atom being in a high oxidation state, then just the opposite occurs. Similar effects occur with electronegative or electropositive elements in the chemical environment of the atom in question. By synthesizing compounds with known structures, patterns can be formed by using XPS and structures of unknown compounds can be determined.

Compound	Peak	Binding energy (eV)
COH	C 1s	286.01 - 286.8
CHF	C 1s	287.5 - 290.2
Nitride	N 1s	396.2 - 398.3
Fe_2O_3	O 1s	529.5 - 530.2
Fe_2O_3	Fe $2p_{3/2}$	710.7 - 710.9
FeO	Fe $2p_{3/2}$	709.1 - 709.5
SiO_2	O 1s	532.5 - 533.3
SiO_2	Si 2p	103.2 - 103.9

Table 13.2: Binding energies of electrons in various compounds.

Sample preparation is important for XPS. Although the technique was originally developed for use with thin, at films, XPS can be used with powders. In order to use XPS with powders, a different method of sample preparation is required. One of the more common methods is to press the powder into a high purity indium foil. A different approach is to dissolve the powder in a quickly evaporating solvent, if possible, which can then be drop-casted onto a substrate. Using sticky carbon tape to adhere the powder to a disc or pressing the sample into a tablet are an option as well. Each of these sample preparations are designed to make the powder compact, as powder not attached to the substrate will contaminate the vacuum chamber. The sample also needs to be completely dry. If it is not, any solvent present in the sample can destroy the necessary high vacuum and contaminate the machine, affecting the data of the current and future samples.

Analyzing functionalized surfaces

Depth profiling

When analyzing a sample by XPS, questions often arise that deal with layers of the sample. For example, is the sample homogenous (Figure 13.2a), with a consistent composition throughout, or layered (Figure 13.2b and c), with certain elements or components residing in specific places in the sample?

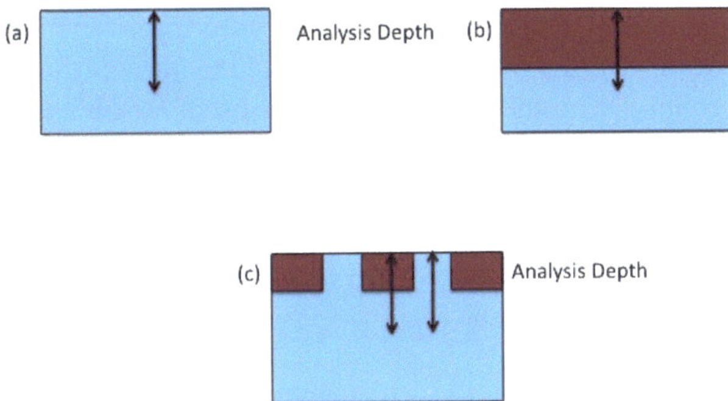

Figure 13.2: Schematic representation of analysis of (a) a homogeneous sample, as compared to (b) a homogeneous layers in a sample, and (c) an inhomogeneous layers in a sample.

A simple way to determine the answer to this question is to perform a depth analysis. By sputtering away the sample, data can be collected at different depths within the sample. It should be noted that sputtering is a destructive process. Within the XPS instrument, the sample is subjected to an Ar^+ ion beam that etches the surface. This creates a hole in the surface, allowing the X-rays to hit layers that would not have otherwise been analyzed. However, it should be realized that different surfaces and layers may be etched at different rates, meaning the same amount of etching does not occur during the same amount of time, depending on the element or compound currently being sputtered.

It is important to note that hydrocarbons sputter very easily and can contaminate the high vacuum of the XPS instrument and thus later samples. They can also migrate to a recently sputtered (and hence unfunctionalized) surface after a short amount of time, so it is imperative to sputter and take a measurement quickly, otherwise the sputtering may appear to have had no effect.

Functionalized films

When running XPS, it is important that the sample is prepared correctly. If it is not, there is a high chance of ruining not only data acquisition, but the instrument as well. With organic functionalization, it is very important to ensure the surface functional group (or as is the case with many functionalized nanoparticles, the surfactant) is immobile on the surface of the substrate. If it is removed easily in the vacuum chamber, it not only will give erroneous data, but it will contaminate the machine, which may then contaminate future samples. This is particularly important when studying thiol functionalization of gold samples, as thiol groups bond strongly with the gold. If there is any loose thiol group contaminating the machine, the thiol will attach itself to any gold sample subsequently placed in the instrument, providing erroneous data. Fortunately, with the above exception, preparing samples that have been functionalized is not much different than standard preparation procedures. However, methods for analysis may have to be modified in order to obtain good, consistent data.

A common method for the analysis of surface modified material is angle resolved X-ray photoelectron spectroscopy (ARXPS). ARXPS is a non-destructive alternative to sputtering, as it relies upon using a series of small angles to analyze the top layer of the sample, giving a better picture of the surface than standard XPS. ARXPS allows for the analysis of the topmost layer of atoms to be analyzed, as opposed to standard XPS, which will analyze a few layers of atoms into the sample, as illustrated in Figure 13.3. ARXPS is often used to analyze surface contaminations, such as oxidation, and surface modification or passivation. Though the methodology and limitations are beyond the scope of this module, it is important to remember that, like normal XPS, ARXPS assumes homogeneous layers are present in samples, which can give erroneous data, should the layers be heterogeneous.

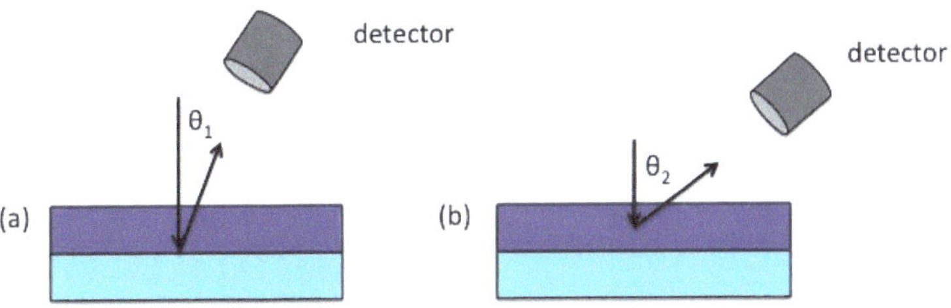

Figure 13.3: Schematic representation of (a) a standard XPS analysis and (b) ARXPS on a multilayer sample.

Limitations of XPS

There are many limitations to XPS that are not based on the samples or preparation, but on the machine itself. One such limitation is that XPS cannot detect hydrogen or helium. This, of course, leads to a ratio of elements in the sample that is not entirely accurate, as there is always some amount of hydrogen. It is a common fallacy to assume the percent of atoms obtained from XPS data are completely accurate due to this presence of undetected hydrogen.

It is possible to indirectly measure the amount of hydrogen in a sample using XPS, but it has to be done in a roundabout, often time-consuming manner. If the sample contains hydrogen with a partial positive charge (e.g., in a metal hydroxide), the sample can be washed in sodium naphthalenide ($C_{10}H_8Na$, Figure 13.4). This replaces this hydrogen with sodium, which can then be measured,

$$[M]\text{-}OH \xrightarrow{C_{10}H_8Na} [M]\text{-}ONa$$

The sodium to oxygen ratio that is obtained infers the hydrogen to oxygen ratio, assuming that all the hydrogen atoms have reacted.

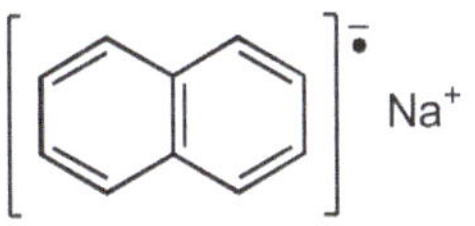

Figure 13.4: The structure of sodium naphthalenide ($C_{10}H_8Na$).

XPS can only give an average measurement, as the electrons lower down in the sample will lose more energy as they pass other atoms while the electrons on the surface retain their original kinetic energy. The electrons from lower layers can also undergo inelastic or elastic scattering, seen in Figure 13.5. This scattering may have a significant impact on data at higher angles of emission. The beam itself is also relatively wide, with the smallest width ranging from 10 - 200 μm, lending to the observed average composition inside the beam area. Due to this, XPS cannot differentiate sections of elements if the sections are smaller than the size of the beam.

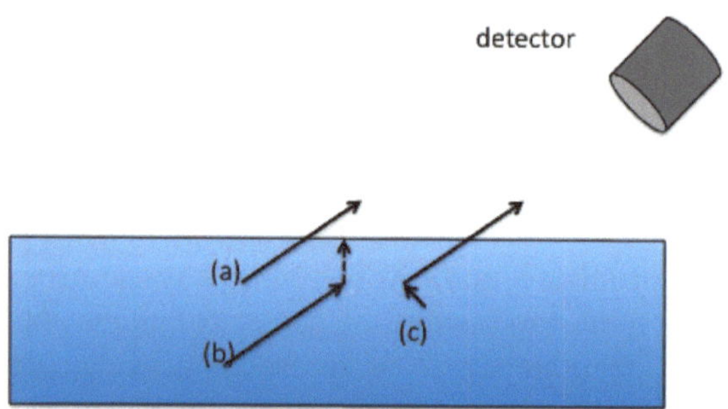

Figure 13.5: Schematic representation of (a) no scattering, (b) inelastic scattering, and (c) elastic scattering.

Sample reaction or degradation are important considerations. Caution should be exercised when analyzing polymers, as they are often chemically active and X-rays will provide energy to start degrading the polymer, altering the properties of the sample. One method found to help overcome this particular limitation is to use angle-resolved X-ray photoelectron spectroscopy (ARXPS). XPS can often reduce certain metal salts, such as Cu^{2+}. This reduction will give peaks that indicate a certain set of properties or chemical environments when it could be completely different.

It needs to be understood that charges can build up on the surface of the sample due to a number of reasons, specifically due to the loss of electrons during the XPS experiment. The charge on the surface will interact with the electrons escaping from the sample, affecting the data obtained. If the charge collecting is positive, the electrons that have been knocked off will be attracted to the charge, slowing the electrons. The detector will pick up a lower kinetic energy of the electrons, and thus calculate a different binding energy than the one expected, giving peaks which could be labeled with an incorrect oxidation state or chemical environment. To overcome this, the spectra must be charge referenced by one of the following methods: using the naturally occurring graphite peak as a reference, sputtering with gold and using the gold peak as a reference or flooding the sample with the ion gun and waiting until the desired peak stops shifting.

Limitations with surfactants and sputtering

While it is known that sputtering is destructive, there are a few other limitations that are not often considered. As mentioned above, the beam of X-rays is relatively large, giving an average composition in the analysis. Sputtering has the same limitation. If the surfactant or layers are not homogeneous, then when the sputtering is finished and detection begins, the analysis will show a homogeneous section, due to the size of both the beam and sputtered area, while it is actually separate sections of elements.

The chemistry of the compounds can be changed with sputtering, as it removes atoms that were bonded, changing the oxidation state of a metal or the hybridization of a non-metal. It can also introduce charges if the sample is non-conducting or supported on a non-conducting surface.

Using XPS to analyze metal nanoparticles

X-ray photoelectron spectroscopy is a surface technique developed for use with thin films. More recently, however, it has been used to analyze the chemical and elemental composition of nanoparticles. The complication of nanoparticles is that they are neither at nor larger than the diameter of the beam, creating issues when using the data obtained at face value. Samples of nanoparticles will often be large aggregates of particles. This creates problems with the analysis acquisition, as there can be a variety of cross-sections, as seen in Figure 13.6. This acquisition problem is also compounded by the fact that the surfactant may not be completely covering the particle, as the curvature of the particle creates defects and divots. Even if it is possible to create a monolayer of particles on a support, other issues are still present. The background support will be analyzed with the particle, due to their small size and the size of the beam and the depth at which it can penetrate.

Many other factors can introduce changes in nanoparticles and their properties. There can be probe, environmental, proximity, and sample preparation effects. The dynamics of particles can wildly vary depending on the reactivity of the particle itself. Sputtering can also be a problem. The beam used to sputter will be roughly the same size or larger than the particles. This means that what appears in the data is not a section of particle, but an average composition of several particles. Each of these issues needs to be taken into account and preventative measures need to be used so the data is the best representation possible.

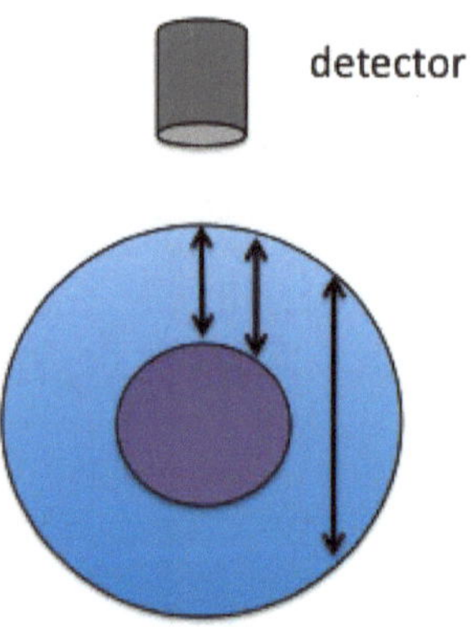

Figure 13.6: Different cross-sections of analysis possible on a nanoparticle.

Sample preparation

Sample preparation of nanoparticles is very important when using XPS. Certain particles, such as iron oxides without surfactants, will interact readily with oxygen in the air. This causes the particles to gain a layer of oxygen contamination. When the particles are then analyzed, oxygen appears where it should not, and the oxidation state of the metal may be changed. As shown by these particles, which call for handling, mounting and analysis without exposure to air, knowing the reactivity of the nanoparticles in the sample is very important even before starting analysis. If the reactivity of the nanoparticle is known, such as the reactivity of oxygen and iron, then preventative steps can be taken in sample preparation in order to obtain the best analysis possible.

When preparing a sample for XPS, a powder form is often used. This preparation, however, will lead to aggregation of nanoparticles. If analysis is performed on such a sample, the data obtained will be an average of composition of each nanoparticle. If composition of a single particle is what is desired, then this average composition will not be sufficient. Fortunately, there are other methods of sample preparation. Samples can be supported on a substrate, which will allow for analysis of single particles. A pictorial representation in Figure 13.7 shows the different types of samples that can occur with nanoparticles.

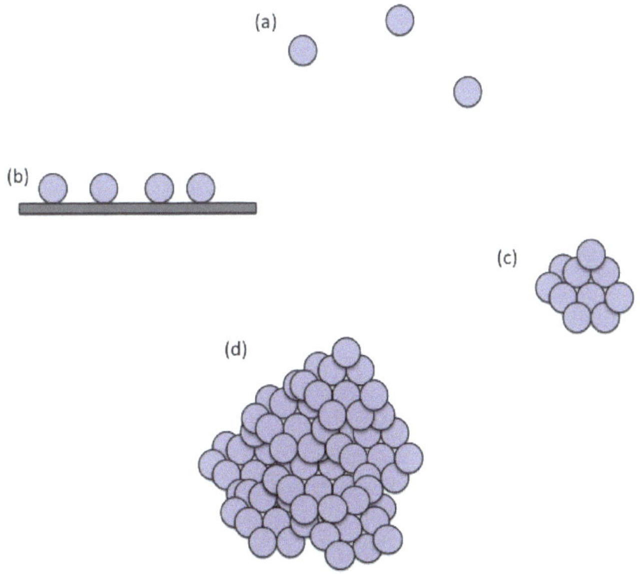

Figure 13.7: Representation of (a) a theoretically isolated nanoparticles, (b) nanoparticles suspended on a substrate, (c) an aggregate of nanoparticles, and (d) a powdered form of nanoparticles.

Analysis limitations

Nanoparticles are dynamic; their properties can change when exposed to new chemical environments, leading to a new set of applications. It is the dynamics of nanoparticles that makes them so useful and is one of the reasons why scientists strive to understand their properties. However, it is this dynamic ability that makes analysis di cult to do properly. Nanoparticles are easily damaged and can change properties over time or with exposure to air, light or any other environment, chemical or otherwise. Surface analysis is often di cult because of the high rate of contamination. Once the particles are inserted into XPS, even more limitations appear.

Probe effects

There are often artifacts introduced from the simple mechanism of conducting the analysis. When XPS is used to analyze the relatively large surface of thin films, there is small change in temperature as energy is transferred. The thin films, however, are large enough that this small change in energy has to significant change to its properties. A nanoparticle is much smaller. Even a small amount of energy can drastically change the shape of particles, in turn changing the properties, giving a much different set of data than expected.

The electron beam itself can affect how the particles are supported on a substrate. Theoretically, nanoparticles would be considered separate from each other and any other chemical environments, such as solvents or substrates. This, however, is not possible, as the particles must be suspended in a solution or placed on a substrate when attempting analysis. The chemical environment around the particle will have some amount of interaction with the particle. This interaction will change characteristics of the nanoparticles, such as oxidation states or partial charges, which will then shift the peaks observed. If particles can be separated and suspended on a substrate, the supporting material will also be analyzed due to the fact that the X-ray beam is larger than the size of each individual particle. If the substrate is made of porous materials, it can adsorb gases and those will be detected along with the substrate and the particle, giving erroneous data.

Environmental effects

Nanoparticles will often react, or at least interact, with their environments. If the particles are highly reactive, there will often be induced charges in the near environment of the particle. Gold nanoparticles have a well-documented ability to undergo plasmon interactions with each other. When XPS is performed on these particles, the charges will change the kinetic energy of the electrons, shifting the apparent binding energy. When working with nanoparticles that are well known for creating charges, it is often best to use an ion gun or a coating of gold. The purpose of the ion gun or gold coating is to try to move peaks back to their appropriate energies. If the peaks do not move, then the chance of there being no induced charge is high and thus the obtained data is fairly reliable.

Proximity effects

The proximity of the particles to each other will cause interactions between the particles. If there is a charge accumulation near one particle, and that particle is in close proximity with other particles, the charge will become enhanced as it spreads, affecting the signal strength and the binding energies of the electrons. While the knowledge of charge enhancement could be useful to potential applications, it is not beneficial if knowledge of the various properties of individual particles is sought.

Less isolated (less crowded) particles will have different properties as compared to more isolated particles. A good example of this is the plasmon effect in gold nanoparticles. The closer gold nanoparticles are to each other, the more likely they will induce the plasmon effect. This can change the properties of the particles, such as oxidation states and partial charges. These changes will then shift peaks seen in XPS spectra. These proximity effects are often introduced in the sample preparation. This, of course, shows why it is important to prepare samples correctly to get desired results.

Summary

Unfortunately, there is no good general procedure for all nanoparticle samples. There are too many variables within each sample to create a basic procedure. A scientist wanting to use XPS to analyze nanoparticles must first understand the drawbacks and limitations of using their sample as well as how to counteract the artifacts that will be introduced in order to properly use XPS.

One must never make the assumption that nanoparticles are at. This assumption will only lead to a misrepresentation of the particles. Once the curvature and stacking of the particles, as well as their interactions with each other are taken into account, XPS can be run.

Bibliography

A. W. Apblett, A. C. Warren, and A. R. Barron, Synthesis and characterization of triethylsiloxy-substituted alumoxanes: their structural relationship to the minerals boehmite and diaspore. *Chem. Mater.*, 1992, **4**, 167.

D. R. Baer and M. H. Engelhard. XPS analysis of nanostructured materials and biological surfaces. *J. Electron Spectrosc.*, 2009, **178-179**, 415.

D. R. Baer, J. E. Amonette, M. H. Engelhard, D. J. Gaspar, A. S. Karakoti, S. Kuchibhatla, P. Nachimuthu, J. T. Nurmi, Y. Qiang, V. Sarathy, S. Seal, A. Sharma. P. G. Tratnyek, and C. M. Wang, Characterization challenges for nanomaterials. *Surf. Interface Anal.*, 2008, **40**, 529.

A. Herrera-Gomez, J. T. Grant, P. J. Cumpson, M. Jenko, F. S. Aguirre-Tostado, C. R. Brundle, T. Conrad, G. Conti, C. S. Fadley, J. Fulghum, K. Kobayashi, L. Kövér, H. Nohira, R. L. Opila, S. Oswald, R. W. Paynter, R. M. Wallace, W. S. M. Werner, and J. Wolstenhome, Report on the 47th IUVSTA workshop 'angle-resolved XPS: the current status and future prospects for angle-resolved XPS of nano and subnano films'. *Surf. Interface Anal.*, 2009, **41**, 840.

C. C. Landry, J. A. Davis, A. W. Apblett, and A. R. Barron, Siloxy-substituted alumoxanes: synthesis from polydialkylsiloxanes and trimethylaluminium, and application as aluminosilicate precursors. *J. Mater. Chem.*, 1993, **3**, 597.

S. F. Mao, Z. M. Zhang, K. Tokesi, A. Csik, J. Toth, R. J. Bereczky, and Z. J. Ding, XPS analysis of nano-thin films on substrate. *Surf. Interface Anal.*, 2008, **40**, 728.

J. F. Moulden, W. F. Stickle, P. E. Sobol, and K. D. Bumben. *Handbook of X-ray Photoelectron Spectroscopy*, Physical Electronics Inc. (1995).

J. P. Sibilia. *Materials Characterization and Chemical Analysis*, 2nd edn., Wiley VCH (1996).

Chapter 14: Auger Electron Spectroscopy

Amanda M. Goodman and Andrew R. Barron

Basic principles

Auger electron spectroscopy (AES) is one of the most commonly employed surface analysis techniques. It uses the energy of emitted electrons to identify the elements present in a sample, similar to X-ray photoelectron spectroscopy (XPS). The main difference is that XPS uses an X-ray beam to eject an electron while AES uses an electron beam to eject an electron. In AES, the sample depth is dependent on the escape energy of the electrons. It is not a function of the excitation source as in XPS. In AES, the collection depth is limited to 1 - 5 nm due to the small escape depth of electrons, which permits analysis of the first 2 - 10 atomic layers. In addition, a typical analysis spot size is roughly 10 nm. A representative AES spectrum illustrating the number of emitted electrons, N, as a function of kinetic energy, E, in direct form (red) and in differentiated form (black) is shown in Figure 14.1.

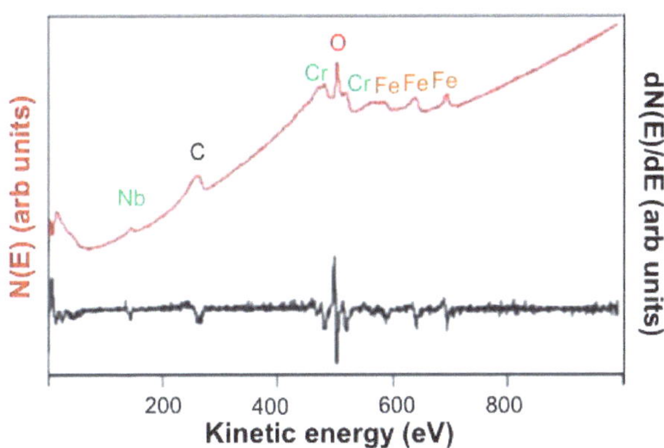

Figure 14.1: AES survey spectrum (red) and differentiated spectrum (black) of an oxidized Fe-Cr-Nb alloy. Adapted from H. J. Mathieu in *Surface Analysis: The Principal Techniques*, Ed. J. C. Vickerman, 2nd edn., Wiley-VCH, Weinheim (2011). Copyright: Wiley-VCH (2011).

Like XPS, AES measures the kinetic energy (E_k) of an electron to determine its binding energy (E_b). The binding energy is inversely proportional to the kinetic energy and can be found from,

$$E_b = h\nu - E_k + \Delta\Phi$$

where $h\nu$ is the energy of the incident photon and $\Delta\Phi$ is the difference in work function between the sample and the detector material.

Since the E_b is dependent on the element and the electronic environment of the nucleus, AES can be used to distinguish elements and their oxidation states. For instance, the energy required to remove an electron from Fe^{3+} is more than in Fe^0. Therefore, the Fe^{3+} peak will have a lower E_k than the Fe^0 peak, effectively distinguishing the oxidation states.

Auger process

An Auger electron comes from a cascade of events. First, an electron beam comes in with sufficient energy to eject a core electron creating a vacancy (A in Figure 14.2). Typical energies of the primary electrons range from 3 - 30 keV. A secondary electron (imaging electron) of higher energy drops down to fill the vacancy (B in Figure 14.2) and emits sufficient energy to eject a tertiary electron (Auger electron) from a higher shell (C in Figure 14.2).

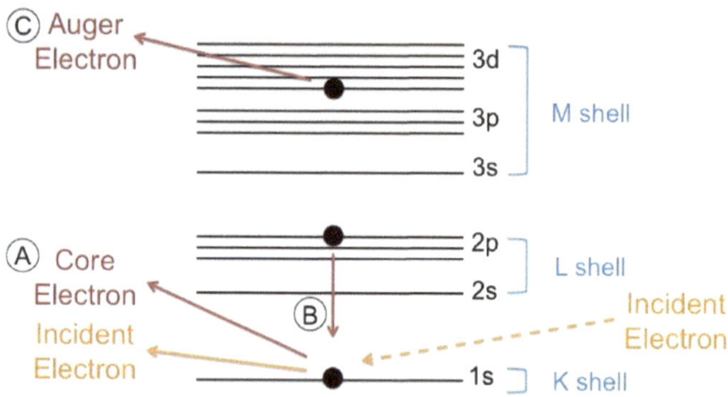

Figure 14.2: Schematic diagram of the Auger process.

The shells from which the electrons move from lowest to highest energy are described as the K shell, L shell, and M shell. This nomenclature is related to quantum numbers. Explicitly, the K shell represents the 1s orbital, the L shell represents the 2s and 2p orbitals, and the M shell represents the 3s, 3p, and 3d orbitals. The cascade of events typically begins with the ionization of a K shell electron, followed by the movement of an L shell electron into the K shell vacancy. Then, either an L shell electron or M shell electron is ejected. It depends on the element which peak is prevalent but often both peaks will be present. The peak seen in the spectrum is labeled according to the shells involved in the movement of the electrons. For example, an electron ejected from a gold atom could be labeled as Au KLL or Au KLM.

The intensity of the peak depends on the amount of material present, while the peak position is element dependent. Auger transitions characteristic of each elements can be found in the literature. Auger transitions of the first forty detectable elements are listed in Table 14.1.

Atomic number	Element	AES transition	Kinetic energy of transition (eV)
3	Li	KLL	43
4	Be	KLL	104
5	B	KLL	179
6	C	KLL	272
7	N	KLL	379
8	O	KLL	508
9	F	KLL	647
11	Na	KLL	990
12	Mg	KLL	1186
13	Al	LMM	68
14	Si	LMM	92
15	P	LMM	120
16	S	LMM	152
17	Cl	LMM	181
19	K	KLL	252
20	Ca	LMM	291
21	Sc	LMM	340
22	Ti	LMM	418
23	V	LMM	473
24	Cr	LMM	529
25	Mn	LMM	589
26	Fe	LMM	703
27	Co	LMM	775
28	Ni	LMM	848
29	Cu	LMM	920
30	Zn	LMM	994
Continued on next page			

Atomic number	Element	AES transition	Kinetic energy of transition (eV)
31	Ga	LMM	1070
32	Ge	LMM	1147
33	As	LMM	1228
34	Se	LMM	1315
35	Br	LMM	1376
39	Y	MNN	127
40	Zr	MNN	147
41	Nb	MNN	167
42	Mo	MNN	186

Table 14.1: Selected AES transitions and their corresponding kinetic energy. Data from H. J. Mathieu in *Surface Analysis: The Principal Techniques*, 2nd edn., Ed. J. C. Vickerman, Wiley-VCH, Weinheim (2011).

Instrumentation

Important elements of an Auger spectrometer include a vacuum system, an electron source, and a detector. AES must be performed at pressures less than 10^{-3} pascal (Pa) to keep residual gases from adsorbing to the sample surface. This can be achieved using an ultra-high-vacuum system with pressures from 10^{-8} to 10^{-9} Pa. Typical electron sources include tungsten laments with an electron beam diameter of 3 - 5 μm, LaB_6 electron sources with a beam diameter of less than 40 nm, and Schottky barrier laments with a 20 nm beam diameter and high beam current density. Two common detectors are the cylindrical mirror analyzer and the concentric hemispherical analyzer discussed below. Notably, concentric hemispherical analyzers typically have better energy resolution.

Cylindrical mirror analyzer (CMA)

A CMA is composed of an electron gun, two cylinders, and an electron detector (Figure 14.3). The operation of a CMA involves an electron gun being directed at the sample. An ejected electron then enters the space between the inner and outer cylinders (IC and OC). The inner cylinder is at ground potential, while the outer cylinder's potential is proportional to the kinetic energy of the electron. Due to its negative potential, the outer cylinder deflects the electron towards the electron detector. Only electrons within the solid angle cone are detected. The resulting signal is proportional to the number of electrons detected as a function of kinetic energy.

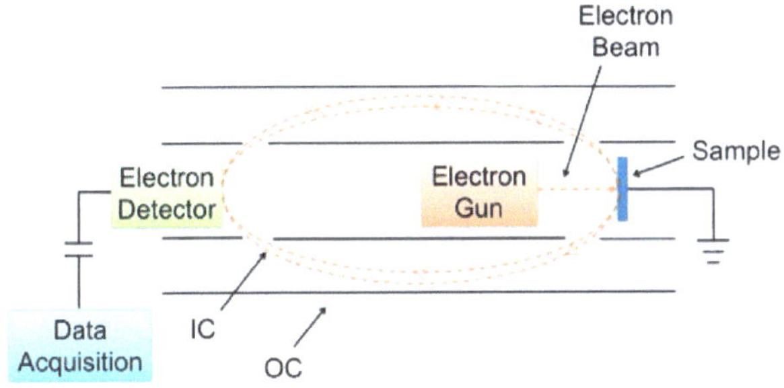

Figure 14.3: Schematic of a cylindrical mirror analyzer.

Concentric hemispherical analyzer (CHA)

A CHA contains three parts (Figure 14.4):

- A retarding and focusing input lens assembly.
- An inner and outer hemisphere (IH and OH).
- An electron detector.

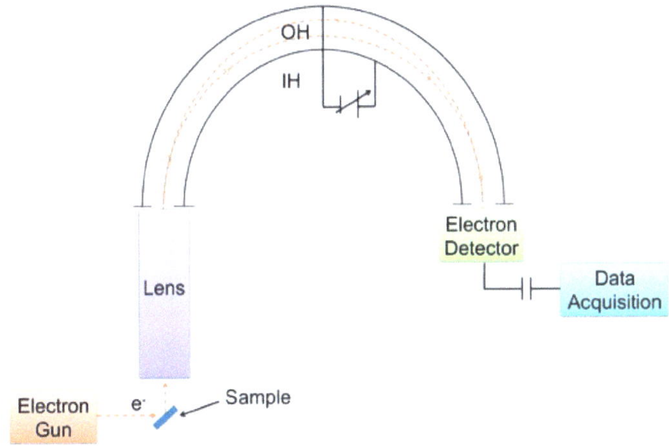

Figure 14.4: Schematic of a concentric hemispherical analyzer.

Electrons ejected from the surface enter the input lens, which focuses the electrons and retards their energy for better resolution. Electrons then enter the hemispheres through an entrance slit. A potential difference is applied on the hemispheres so that only electrons with a small range of energy differences reach the exit. Finally, an electron detector analyzes the electrons.

Applications

AES has widespread use owing to its ability to analyze small spot sizes with diameters from 5 μm down to 10 nm depending on the electron gun. For instance, AES is commonly employed to study film growth and surface-chemical composition, as well as grain boundaries in metals and ceramics. It is also used for quality control surface analyses in integrated circuit production lines due to short acquisition times. Moreover, AES is used for areas that require high spatial resolution, which XPS cannot achieve. AES can also be used in conjunction with transmission electron microscopy (TEM) and scanning electron microscopy (SEM) to obtain a comprehensive understanding of microscale materials, both chemically and structurally. As an example of combining techniques to investigate microscale materials, Figure 14.5 shows the characterization of a single wire from a Sn-Nb multi-wire alloy. Figure 14.5a is a SEM image of the singular wire and Figure 14.5b is a schematic depicting the distribution of Nb and Sn within the wire. Point analysis was performed along the length of the wire to determine the percent concentrations of Nb and Sn.

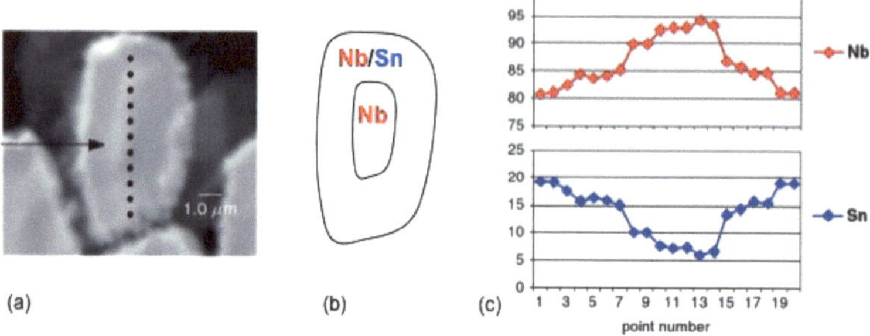

Figure 14.5: Analysis of a Sn-Nb wire. (a) SEM image of the wire, (b) schematic of the elemental distribution, and (c) graphical representation of point analysis giving the percent concentration of Nb and Sn. Adapted from H. J. Mathieu in *Surface Analysis: The Principal Techniques*, 2nd edn., Ed. J. C. Vickerman, Wiley-VCH, Weinheim (2011). Copyright: Wiley-VCH (2011).

AES is widely used for depth profiling. Depth profiling allows the elemental distributions of layered samples 0.2 1 μm thick to be characterized beyond the escape depth limit of an electron. Varying the incident and collection angles, and the primary beam energy controls the analysis depth. In general, the depth resolution decreases with the square root of the sample thickness. Notably, in AES, it is possible to simultaneously sputter and collect Auger data for depth profiling. The sputtering time indicates the depth and the intensity indicates elemental concentrations. Since, the sputtering process does not affect the ejection of the Auger electron, helium or argon ions can be used to sputter the surface and create the trench, while collecting Auger data at the same time. The depth profile does not have the problem of diffusion of hydrocarbons into the trenches. Thus, AES is better for depth profiles of reactive metals (e.g., gold or any metal or semiconductor). Yet, care should be taken because sputtering can mix up different elements, changing the sample composition.

Limitations

While AES is a very valuable surface analysis technique, there are limitations. Because AES is a three-electron process, elements with less than three electrons cannot be analyzed. Therefore, hydrogen and helium cannot be detected. Nonetheless, detection is better for lighter elements with fewer transitions. The numerous transition peaks in heavier elements can cause peak overlap, as can the increased peak width of higher energy transitions. Detection limits of AES include 0.1 1% of a monolayer, 10^{-16} - 10^{-15} g of material, and 10^{10} - 10^{13} atoms/cm^2.

Another limitation is sample destruction. Although focusing of the electron beam can improve resolution; the high-energy electrons can destroy the sample. To limit destruction, beam current densities of greater than 1 mA/cm^2 should be used. Furthermore, charging of the electron beam on insulating samples can deteriorate the sample and result in high-energy peak shifts or the appearance of large peaks.

Bibliography

H. Bubert, J. Rivière, and W. Werner, Surface and Thin Film Analysis: *A Compendium of Principles, Instrumentation, and Applications*, 2nd edn., Ed. G. Friedbacher and H. Bubert, Wiley-VCH, Weinheim (2011).

H. Lüth in Solid Surfaces, *Interfaces and Thin Films*, 5th edn., Springer, New York (2010).

H. J. Mathieu in *Surface Analysis: The Principal Techniques*, 2nd edn., Ed. J. C. Vickerman, Wiley-VCH, Weinheim (2011).

S. N. Raman, D. F. Paul, J. S. Hammond, and K. D. Bomben, Auger electron spectroscopy and its application to nanotechnology. *Microscopy Today*, 2011, **19**, 12.

N. Turner in *Analytical Instrumentation Handbook*, Second Edition, Ed. G. Ewing, Marcel Dekker, Inc, New York (1997).

J. F. Watts in *Handbook of Adhesion Technology*, Ed. Lucas F.M. da Silva, A. Öchsner, and R. Adams, Springer, New York (2011).

V. Young and G. Hoflund in *Handbook of Surface and Interface Analysis Methods for Problem-Solving*, 2nd edn., Ed. J. Rivière and S. Myhra, CRC Press, Boca Raton (2009).

Chapter 15: Rutherford Backscattering

Avishek Saha and Andrew R. Barron

Introduction

One of the main research interests of the semiconductor industry is to improve the performance of semiconducting devices and to construct new materials with reduced size or thickness that have potential application in transistors and microelectronic devices. However, the most significant challenge regarding thin film semiconductor materials is measurement. Properties such as the thickness, composition at the surface, and contamination, all are critical parameters of the thin films. To address these issues, we need an analytical technique which can measure accurately through the depth of the of the semiconductor surface without destruction of the material. Rutherford backscattering spectroscopy is a unique analysis method for this purpose. It can give us information regarding in-depth profiling in a non-destructive manner. X-ray photo electron spectroscopy (XPS), energy dispersive X-ray analysis (EDX) and Auger electron spectroscopy are also able to study the depth-profile of semiconductor films. Table 15.1 demonstrates the comparison between those techniques with RBS.

Method	Destructive	Incident particle	Outgoing Particle	Detection limit	Depth resolution
RBS	No	Ion	Ion	~1	10 nm
XPS	Yes	X-ray photon	Electron	~0.1-1	~1 μm
EDX	Yes	Electron	X-ray photon	~0.1	1.5 nm
Auger	Yes	Electron	Electron	~0.1-1	1.5 nm

Table 15.1: Comparison between different thin film analysis techniques.

Basic concept

At a basic level, RBS demonstrates the electrostatic repulsion between high energy incident ions and target nuclei. Rutherford's (Figure 15.1) gold foil experiments demonstrated that a small portion of alpha particles ($^4He^+$) were deflected by a gold foil due to the presence of a small high-density nucleus (Figure 15.2a). It was this unexpected result that led to Rutherford's model for the atom that included a small, dense, nucleus, and it is this backscattering of an alpha particle that is the basis for RBS.

Figure 15.1: New Zealand-born British physicist Ernest Rutherford, 1st Baron Rutherford of Nelson, FRS (1871 - 1937), who received the Nobel prize in 1908 for "his investigations into the disintegration of the elements, and the chemistry of radioactive substances", but is more famous for the model of atomic structure.

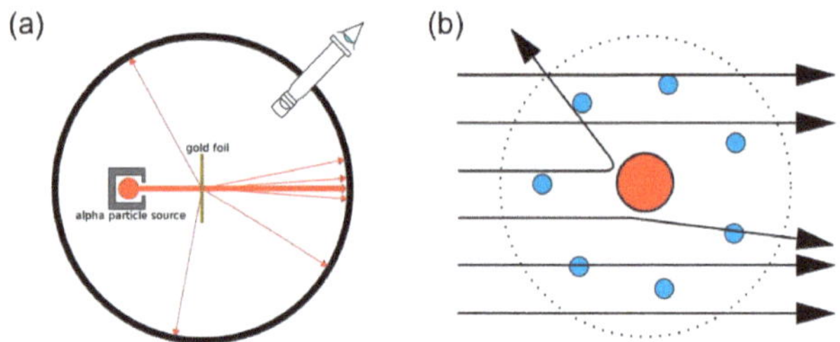

Figure 15.2: (a) The observed results of Rutherford's gold foil experiment where a small portion of the alpha particles were deflected, indicating a small, concentrated positive charge, that led to Rutherford's model of the atom (b).

In RBS the specimen under study is bombarded with monoenergetic beam of $^4\text{He}^+$ particles and the backscattered particles are detected by the detector-analysis system which measures the energies of the particles. During the collision, energy is transferred from the incident particle to the target specimen atoms; the change in energy of the scattered particle depends on the masses of incoming and target atoms. For an incident particle of mass M_1, the energy is E_0 while the mass of the target atom is M_2. After the collision, the residual energy E of the particle scattered at angle Ø can be expressed as:

$$E - k^2 E_0$$

$$k = \frac{M_1\cos\phi + \sqrt{(M_2{}^2 - M_1{}^2\sin 2\ \phi)}}{M_1 + M_2}$$

where k is the kinematic scattering factor, which is actually the energy ratio of the particle before and after the collision. Since k depends on the masses of the incident particle and target atom and the scattering angle, the energy of the scattered particle is also determined by these three parameters. A simplified layout of backscattering experiment is shown in Figure 15.3.

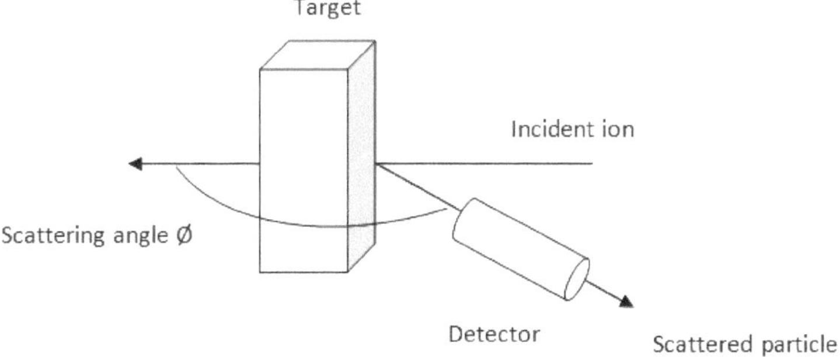

Figure 15.3: Schematic representation of the experimental setup for Rutherford backscattering analysis.

The probability of a scattering event can be described by the differential scattering cross section of a target atom for scattering an incoming particle through the angle Ø into differential solid angle as follows,

$$\frac{d\sigma_R}{d\varphi} = \left(\frac{zZ^2}{2E_0\sin^2\phi}\right)\frac{[\cos\phi + \sqrt{1 - (M_1/M_2\sin\phi)^2}]^2}{\sqrt{1 - (M_1/M_2\sin\phi)]^2}}$$

where $d\sigma_R$ is the effective differential cross section for the scattering of a particle. The above equation may look complicated, but it conveys the message that the probability of scattering event can be expressed as a function of scattering cross section which is proportional to the zZ when a particle with charge ze approaches the target atom with charge Ze.

Helium ions not scattered at the surface lose energy as they traverse the solid. They lose energy due to interaction with electrons in the target. After collision the He particles lose further energy on their way out to the detector. We need to know two quantities to measure the energy loss, the distance Δt that the particles penetrate into the target and the energy loss ΔE in this distance Figure 15.4. The rate of energy loss or stopping power is a critical component in backscattering experiments as it determines the depth profile in a given experiment.

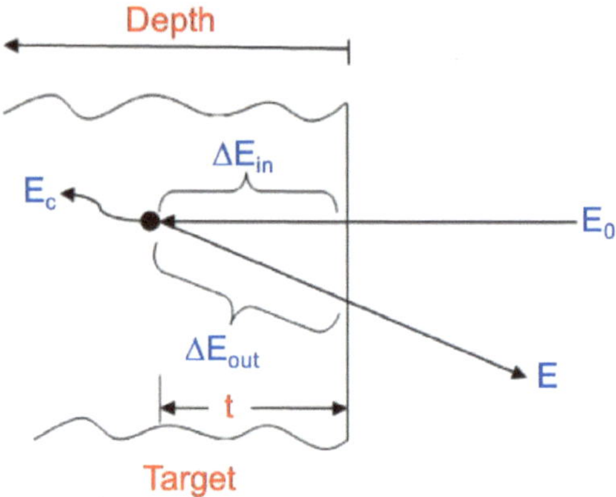

Figure 15.4: Components of energy loss for an ion beam that scatters from depth (t). First, incident beam loses energy through interaction with electrons ΔE_{in}. Then energy lost occurs due to scattering E_c. Finally, the outgoing beam loses energy for interaction with electrons ΔE_{out}. Adapted from L. C. Feldman and J. W. Mayer, *Fundamentals of Surface and Thin Film Analysis*, North Holland-Elsevier, New York (1986).

In thin film analysis, it is convenient to assume that total energy loss ΔE into depth, t, is only proportional to depth for a given target. This assumption allows a simple derivation of energy loss in backscattering as more complete analysis requires many numerical techniques. In constant dE/dx approximation, total energy loss becomes linearly related to depth fit, Figure 15.5.

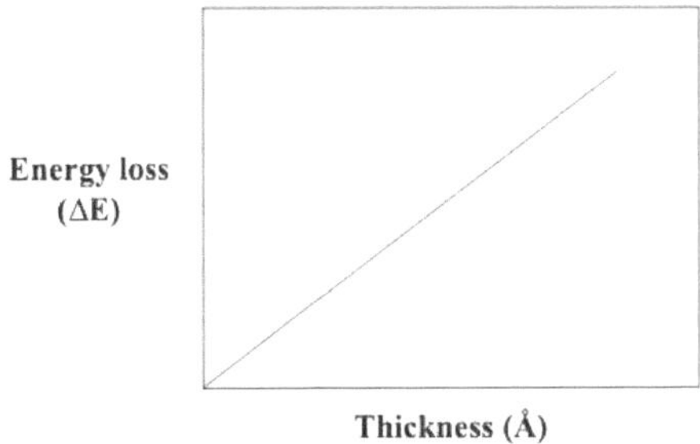

Figure 15.5: Variation of energy loss with the depth of the target in constant dE/dx approximation.

Experimental set-up

The apparatus for Rutherford backscattering analysis of thin solid surface typically consist of three components:

- A source of helium ions.
- An accelerator to energize the helium ions.
- A detector to measure the energy of scattered ions.

There are two types of accelerator/ion source available. In single stage accelerator, the He^+ source is placed within an insulating gas-filled tank (Figure 15.6). It is difficult to install new ion source when it is exhausted in this type of accelerator. Moreover, it is also difficult to achieve particles with energy much more than 1 MeV since it is difficult to apply high voltages in this type of system.

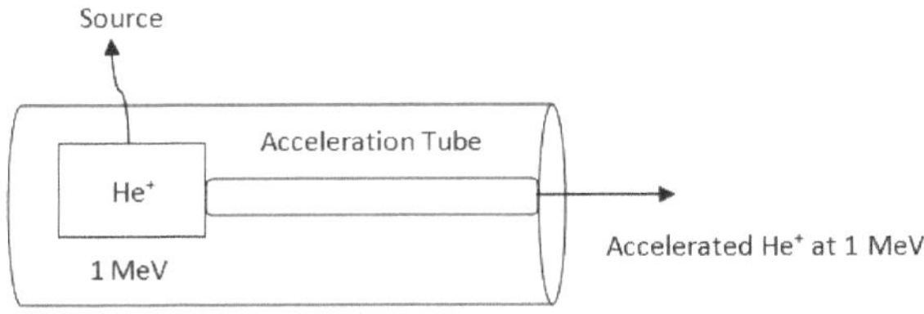

Figure 15.6: Schematic representation of a tandem accelerator.

Another variation is tandem accelerator. Here the ion source is at ground and produces negative ion. The positive terminal is located is at the center of the acceleration tube (Figure 15.7). Initially the negative ion is accelerated from ground to terminal. At terminal two-electron stripping process converts the He^- to He^{++}. The positive ions are further accelerated toward ground due to columbic repulsion from positive terminal. This arrangement can achieve highly accelerated He^{++} ions ($\sim$2.25 MeV) with moderate voltage of 750 kV.

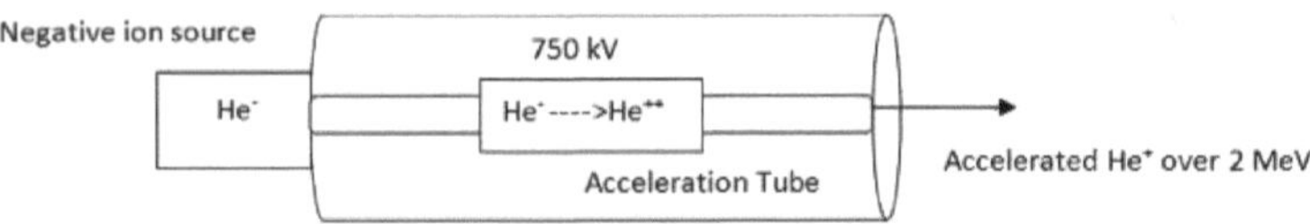

Figure 15.7: Schematic representation of a single stage accelerator.

Particles that are backscattered by surface atoms of the bombarded specimen are detected by a surface barrier detector. The surface barrier detector is a thin layer of p-type silicon on the n-type substrate resulting p-n junction. When the scattered ions exchange energy with the electrons on the surface of the detector upon reaching the detector, electrons get promoted from the valence band to the conduction band. Thus, each exchange of energy creates electron-hole pairs. The energy of scattered ions is detected by simply counting the number of electron-hole pairs. The energy resolution of the surface barrier detector in a standard RBS experiment is 12 - 20 keV. The surface barrier detector is generally set between 90° and 170° to the incident beam. Films are usually set normal to the incident beam. A simple layout is shown in Figure 15.8.

Depth profile analysis

As stated earlier, it is a good approximation in thin film analysis that the total energy loss ΔE is proportional to depth t. With this approximation, we can derive the relation between energy width ΔE of the signal from a film of thickness Δt as follows,

$$\Delta E = \Delta t(k\ dE/dx_{in} + 1/cos\emptyset\ dE/dx_{out})$$

where $\emptyset$ = lab scattering angle. It is worth noting that k is the kinematic factor defined in equation above and the subscripts in and out indicate the energies at which the rate of loss of energy or dE/dx is evaluated. As an example, we consider the backscattering spectrum, at scattering angle 170°, for 2 MeV

He^{++} incidents on silicon layer deposited onto 2 mm thick niobium substrate Figure 15.9.

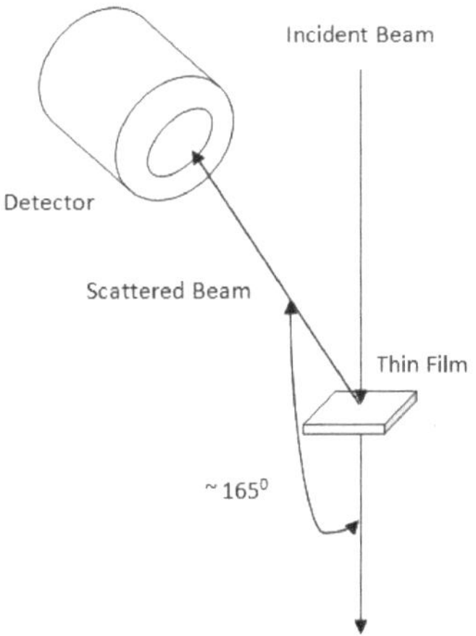

Figure 15.8: Schematic representation general setup where the surface barrier detector is placed at angle of 165° to the extrapolated incident beam.

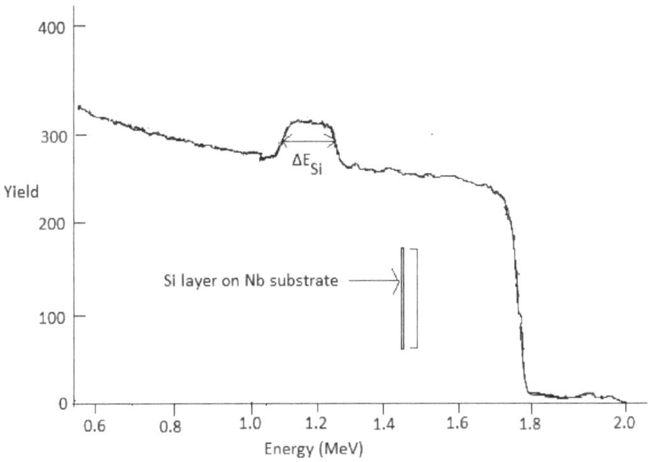

Figure 15.9: The backscattering spectrum for 2.0 MeV He ions incident on a silicon thin film deposited onto a niobium substrate. Adapted from P. D. Stupik, M. M. Donovan, A. R. Barron, T. R. Jervis and M. Nastasi, The interfacial mixing of silicon coatings on niobium metal: a comparative study. *Thin Solid Films*, 1992, 207, 138. Copyright: Elsevier (1992).

The energy loss rate of incoming He^{++} or dE/dx along inward path in elemental Si is $\approx$24.6 eV/Å at 2 MeV and is $\approx$26 eV/Å for the outgoing particle at 1.12 MeV (Since K of Si is 0.56 when the scattering angle is 170° , energy of the outgoing particle would be equal to 2 × 0.56 or 1.12 MeV). Again, the value of ΔE_{Si} is $\approx$133.3 keV. Putting the values into above equation we get

$$\Delta t \approx 133.3 \text{ keV}/(0.56 \times 24.6 \text{ eV/Å} + 1/\cos 170° \times 26 \text{ eV/Å})$$

$$\Delta t = 133.3 \text{ keV}/(13.77 \text{ eV/Å} + 29/.985 \text{ eV/Å})$$

$$\Delta t = 133.3 \text{ keV}/ 40.17 \text{ eV/Å}$$

$$\Delta t = 3318 \text{ Å}$$

Hence a Si layer of ca. 3300 Å thickness has been deposited on the niobium substrate. However, we need to remember that the value of dE/dx is approximated in this calculation.

Quantitative Analysis

In addition to depth pro le analysis, we can study the composition of an element quantitatively by backscattering spectroscopy. The basic equation for quantitative analysis is,

$$Y = \sigma.\Omega.Q.N\Delta t$$

where Y is the yield of scattered ions from a thin layer of thickness Δt, Q is the number of incident ions and Ω is the detector solid angle, and $N\Delta t$ is the number of specimen atoms (atom/cm^2). Figure 15.10 shows the RBS spectrum for a sample of silicon deposited on a niobium substrate and subjected to laser mixing. The Nb has reacted with the silicon to form a $NbSi_2$ interphase layer. The Nb signal has broadened after the reaction.

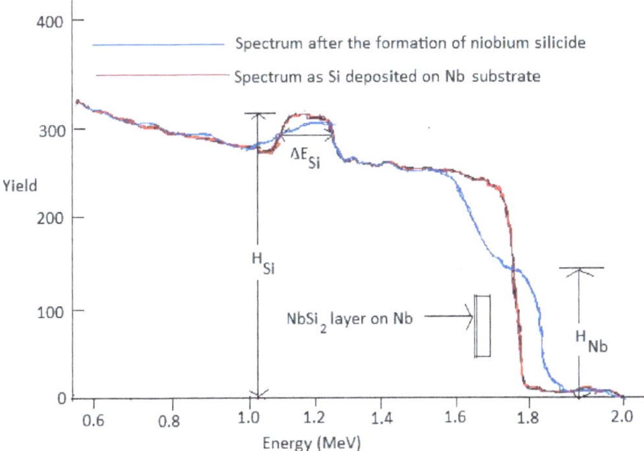

Figure 15.10: Backscattering spectra of Si di used into Nb and Si as deposited on Nb substrate. Adapted from P. D. Stupik, M. M. Donovan, A. R. Barron, T. R. Jervis and M. Nastasi, The interfacial mixing of silicon coatings on niobium metal: a comparative study. *Thin Solid Films*, 1992, 207, 138. Copyright: Elsevier (1992).

We can use ratio of the heights H_{Si}/H_{Nb} of the backscattering spectrum after formation of $NbSi_2$ to determine the composition of the silicide layer. The stoichiometric ratio of Nb and Si can be approximated as,

$$N_{Si}/N_{Nb} \approx [H_{Si} \times \sigma_{Si}]/[H_{Nb} \times \sigma_{Nb}]$$

Hence the concentration of Si and Nb can be determined if we can know the appropriate cross sections σ_{Si} and σ_{Nb}. However, the yield in the backscattering spectra is better represented as the product of signal height and the energy width ΔE. Thus, stoichiometric ratio can be better approximated as

$$N_{Si}/N_{Nb} \approx [H_{Si} \times \Delta E_{Si} \times \sigma_{Si}]/[H_{Nb} \times \Delta E_{Nb} \times \sigma_{Nb}]$$

Limitations

It is of interest to understand the limitations of the backscattering technique in terms of the comparison with other thin film analysis technique such as AES, XPS and SIMS (Table 15.1). AES has better mass resolution, lateral resolution and depth resolution than RBS. But AES suffers from sputtering artifacts. Compared to RBS, SIMS has better sensitivity. RBS does not provide any chemical bonding information which we can get from XPS. Again, sputtering artifact problems are also associated in XPS. The strength of RBS lies in quantitative analysis. However, conventional RBS systems cannot analyze ultrathin films since the depth resolution is only about 10 nm using surface barrier detector.

Summary

Rutherford Backscattering analysis is a straightforward technique to determine the thickness and composition of thin films (< 4000 Å). Areas that have been lately explored are the use of backscattering technique in composition determination of new superconductor oxides; analysis of lattice mismatched epitaxial layers, and as a probe of thin film morphology and surface clustering.

Bibliography

L. C. Feldman and J. W. Mayer, *Fundamentals of Surface and Thin Film Analysis*, North Holland-Elsevier, New York (1986).

A. N. MacInnes and A. R. Barron, A spectroscopic evaluation of the efficacy of two mass deacidification processes for paper. *J. Mater. Chem.*, 1992, **2**, 1049.

Ion Spectroscopies for Surface Analysis, Ed. A. W. Czanderna and D. M. Hercules, Plenum Press (New York), 1991.

P. D. Stupik, M. M. Donovan, A. R. Barron, T. R. Jervis and M. Nastasi, The interfacial mixing of silicon coatings on niobium metal: a comparative study. *Thin Solid Films*, 1992, **207**, 138.

Chapter 16: Gamma-ray Spectroscopy

Brandon Cisneros and Andrew R. Barron

Introduction

Gamma-ray (γ-ray) spectroscopy is a quick and nondestructive analytical technique that can be used to identify various radioactive isotopes in a sample. In gamma-ray spectroscopy, the energy of incident gamma-rays is measured by a detector. By comparing the measured energy to the known energy of gamma-rays produced by radioisotopes, the identity of the emitter can be determined. This technique has many applications, particularly in situations where rapid nondestructive analysis is required.

Background principles

Radioactive decay

The field of chemistry typically concerns itself with the behavior and interactions of stable isotopes of the elements. However, elements can exist in numerous states which are not stable. For example, a nucleus can have too many neutrons for the number of protons it has or contrarily, it can have too few neutrons for the number of protons it has. Alternatively, the nuclei can exist in an excited state, wherein a nucleon is present in an energy state that is higher than the ground state. In all of these cases, the unstable state is at a higher energy state and the nucleus must undergo some kind of decay process to reduce that energy.

There are many types of radioactive decay but type most relevant to gamma-ray spectroscopy is gamma decay. When a nucleus undergoes radioactive decay by α or β decay, the resultant nucleus produced by this process, often called the daughter nucleus, is frequently in an excited state. Similar to how electrons are found in discrete energy levels around a nucleus, nucleons are found in discrete energy levels within the nucleus. In γ decay, the excited nucleon decays to a lower energy state and the energy difference is emitted as a quantized photon. Because nuclear energy levels are discrete, the transitions between energy levels are fixed for a given transition. The photon emitted from a nuclear transition is known as a γ-ray.

Radioactive decay kinetics and equilibria

Radioactive decay, with few exceptions, is independent of the physical conditions surrounding the radioisotope. As a result, the probability of decay at any given instant is constant for any given nucleus of that particular radioisotope. We can use calculus to see how the number of parent nuclei present varies with time. The time constant, λ, is a representation of the rate of decay for a given nuclei,

$$dN/N = -\lambda dt$$

If the symbol N_0 is used to represent the number of radioactive nuclei present at $t = 0$, then the following equation describes the number of nuclei present at some given time.

$$N = N_0 e^{-\lambda t}$$

The same equation can be applied to the measurement of radiation with some sort of detector. The count rate will decrease from some initial count rate in the same manner that the number of nuclei will decrease from some initial number of nuclei.

The decay rate can also be represented in a way that is more easily understood. The equation describing half-life ($t_{1/2}$) is shown in

$$t_{1/2} = \ln2/\lambda$$

The half-life has units of time and is a measure of how long it takes for the number of radioactive nuclei in a given sample to decrease to half of the initial quantity. It provides a conceptually easy way to compare the decay rates of two radioisotopes. If one has a the same number of starting nuclei for two radioisotopes, one with a short half-life and one with a long half-life, then the count rate will be higher for the radioisotope with the short half-life, as many more decay events must happen per unit time in order for the half-life to be shorter.

When a radioisotope decays, the daughter product can also be radioactive. Depending upon the relative half-lives of the parent and daughter, several situations can arise: no equilibrium, a transient equilibrium, or a secular equilibrium. This module will not discuss the former two possibilities, as they are of less relevance to this particular discussion.

Secular equilibrium takes place when the half-life of the parent is much longer than the half-life of the daughter. In any arbitrary equilibrium, the ratio of atoms of each can be described as in,

$$N_p/N_D = (\lambda_D - \lambda_p)/\lambda_p$$

Because the half-life of the parent is much, much greater than the daughter, as the parent decays, the observed amount of activity changes very little.

$$N_p/N_D = \lambda_D/\lambda_p$$

This can be rearranged to show that the activity of the daughter should equal the activity of the parent.

$$A_p = A_D$$

Once this point is reached, the parent and the daughter are now in secular equilibrium with one another and the ratio of their activities should be fixed. One particularly useful application of this concept, to be discussed in more detail later, is in the analysis of the refinement level of long-lived radioisotopes that are relevant to trafficking.

Detectors

Scintillation detector

A scintillation detector is one of several possible methods for detecting ionizing radiation (Figure 16.1). Scintillation is the process by which some material, be it a solid, liquid, or gas, emits light in response to incident ionizing radiation. In practice, this is used in the form of a single crystal of sodium iodide that is doped with a small amount of thallium, referred to as NaI(Tl). This crystal is coupled to a photomultiplier tube which converts the small ash of light into an electrical signal through the photoelectric effect. This electrical signal can then be detected by a computer.

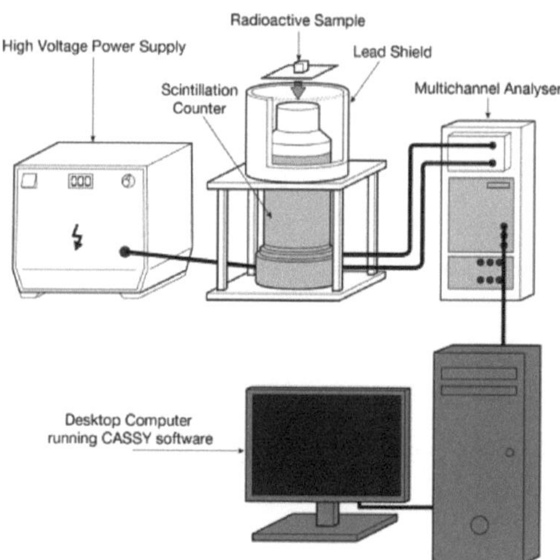

Figure 16.1. Laboratory equipment for determination of γ-radiation spectrum with a scintillation counter. The output from the scintillation counter goes to a Multichannel Analyzer which processes and formats the data.

Semiconductor detector

A semiconductor accomplishes the same effect as a scintillation detector, conversion of gamma radiation into electrical pulses, except through a different route. In a semiconductor, there is a small energy gap between the valence band of electrons and the conduction band. When a semiconductor is hit with gamma-rays, the energy imparted is sufficient to promote electrons to the conduction band. This change in conductivity can be detected and a signal can be generated correspondingly. Germanium crystals doped with lithium, Ge(Li), and high-purity germanium (HPGe) detectors are among the most common types.

Advantages and disadvantages

Each detector type has its own advantages and disadvantages. The NaI(Tl) detectors are generally inferior to Ge(Li) or HPGe detectors in many respects, but are superior to Ge(Li) or HPGe detectors in cost, ease of use, and durability. Germanium-based detectors generally have much higher resolution than NaI(Tl) detectors. Many small photopeaks are completely undetectable on NaI(Tl) detectors that are plainly visible on germanium detectors. However, Ge(Li) detectors must be kept at cryogenic temperatures for the entirety of their lifetime or else they rapidly because incapable of functioning as a gamma-ray detector. Sodium iodide detectors are much more portable and can even potentially be used in the field because they do not require cryogenic temperatures so long as the photopeak that is being investigated can be resolved from the surrounding peaks.

Gamma spectrum features

There are several dominant features that can be observed in a gamma spectrum. The dominant feature that will be seen is the photopeak. The photopeak is the peak that is generated when a gamma-ray is totally absorbed by the detector. Higher density detectors and larger detector sizes increase the probability of the gamma-ray being absorbed.

The second major feature that will be observed is that of the Compton edge and distribution. The *Compton edge* arises due to *Compton effect*, wherein a portion of the energy of the gamma-ray is transferred to the semiconductor detector or the scintillator. This occurs when the relatively high energy gamma ray strikes a relatively low energy electron. There is a relatively sharp edge to the Compton edge that corresponds to the maximum amount of energy that can be transferred to the electron via this type of scattering. The broad peak lower in energy than the Compton edge is the *Compton distribution* and corresponds to the energies that result from a variety of scattering angles. A feature in Compton distribution is the backscatter peak. This peak is a result of the same effect but corresponds to the minimum energy amount of energy transferred. The sum of the energies of the Compton edge and the backscatter peak should yield the energy of the photopeak.

Another group of features in a gamma spectrum are the peaks that are associated with pair production. Pair production is the process by which a gamma ray of sufficiently high energy (>1.022 MeV) can produce an electron-positron pair. The electron and positron can annihilate and produce two 0.511 MeV gamma photons. If all three gamma rays, the original with its energy reduced by 1.022 MeV and the two annihilation gamma rays, are detected simultaneously, then a full energy peak is observed. If one of the annihilation gamma rays is not absorbed by the detector, then a peak that is equal to the full energy less 0.511 MeV is observed. This is known as an escape peak. If both annihilation gamma rays escape, then a full energy peak less 1.022 MeV is observed. This is known as a double escape peak.

Example of experiments

Determination of depleted uranium

Natural uranium is composed mostly of ^{238}U with low levels of ^{235}U and ^{234}U. In the process of making enriched uranium, uranium with a higher level of ^{235}U, depleted uranium is produced. Depleted uranium is used in many applications particularly for its high density. Unfortunately, uranium is toxic and is a potential health hazard and is sometimes found in trafficked radioactive materials, so it is important to have a methodology for detection and analysis of it.

One easy method for this determination is achieved by examining the spectrum of the sample and comparing it qualitatively to the spectrum of a sample that is known to be natural uranium. This type of qualitative approach is not suitable for issues that are of concern to national security. Fortunately, the same approach can be used in a quantitative fashion by examining the ratios of various gamma-ray photopeaks.

The concept of a radioactive decay chain is important in this determination. In the case of ^{238}U, it decays over many steps to ^{206}Pb. In the process, it goes through ^{234m}Pa, ^{234}Pa, and ^{234}Th. These three isotopes have detectable gamma emissions that are capable of being used quantitatively (Figure 16.2). As can be seen in Table 16.1, the half-life of these three emitters is much less than the half-life of ^{238}U. As a result, these should exist in secular equilibrium with ^{238}U. Given this, the ratio of activity of ^{238}U to each daughter products should be 1:1. They can thus be used as a surrogate for measuring ^{238}U decay directly via gamma spectroscopy. The total activity of the ^{238}U can be determined by,

$$A = R/B$$

where A is the total activity of ^{238}U, R is the count rate of the given daughter isotope, and B is the probability of decay via that mode. The count rate may need to be corrected for self-absorption of the sample is particularly thick. It may also need to be corrected for detector efficiency if the instrument does not have some sort of internal calibration.

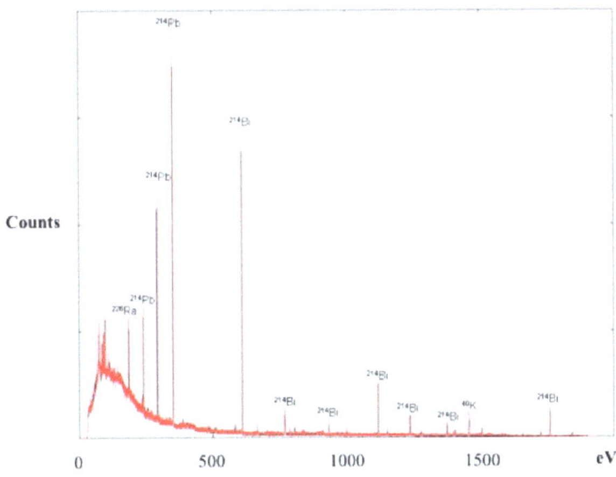

Figure 16.2. The gamma-ray spectrum of natural uranium, showing about a dozen discrete lines superimposed on a smooth continuum, allows one to identify the nuclides ^{226}Ra, ^{214}Pb, and ^{214}Bi of the uranium decay chain.

Isotope	Half-life
^{238}U	4.5×10^9 years
^{234}Th	24.1 days
^{234m}Pa	1.17 minutes

Table 16.1: Half-lives of pertinent radioisotopes in the ^{238}U decay chain.

The count rate of ^{235}U can be observed directly with gamma spectroscopy. This can be converted, as was done in the case of ^{238}U above, to the total activity of ^{235}U present in the sample. Given that the natural abundances of ^{238}U and ^{235}U are known, the ratio of the expected activity of ^{238}U to ^{235}U can be calculated to be a ratio of 21.72:1. If the calculated ratio of disintegration rates varies significantly from this expected value, then the sample can be determined to be depleted or enriched.

This type of calculation is not unique to ^{238}U. It can be used in any circumstance where the ratio of two isotopes needs to be compared so long as the isotope itself or a daughter product it is in secular equilibrium with has a usable gamma-ray photopeak.

Determination of the age of highly enriched uranium

Particularly in the investigation of trafficked radioactive materials, particularly fissile materials, it is of interest to determine how long it has been since the sample was enriched. This can help provide an idea of the source of the fissile material if it was enriched for the purpose of trade or if it was from cold war era enrichment, etc.

When uranium is enriched, ^{235}U is concentrated in the enriched sample by removing it from natural uranium. This process will separate the uranium from its daughter products that it was in secular equilibrium with. In addition, when ^{235}U is concentrated in the sample, ^{234}U is also concentrated due to the particulars of the enrichment process. The ^{234}U that ends up in the enriched sample will decay through several intermediates to ^{214}Bi. By comparing the activities of ^{234}U and ^{214}Bi or ^{226}Ra, the age of the sample can be determined.

$$A_{Bi} = A_{Ra} = \frac{A_U}{2} \lambda_{Th}\lambda_{Ra}T^2$$

where A_{Bi} is the activity of ^{214}Bi, A_{Ra} is the activity of ^{226}Ra, A_U is the activity of ^{234}U, λ_{Th} is the decay constant for ^{230}Th, λ_{Ra} is the decay constant for ^{226}Ra, and T is the age of the sample. This is a simplified form of a more complicated equation that holds true over all practical sample ages (on the order of years) due to the very long half-lives of the isotopes in question. The results of this can be graphically plotted as they are in Figure 16.3.

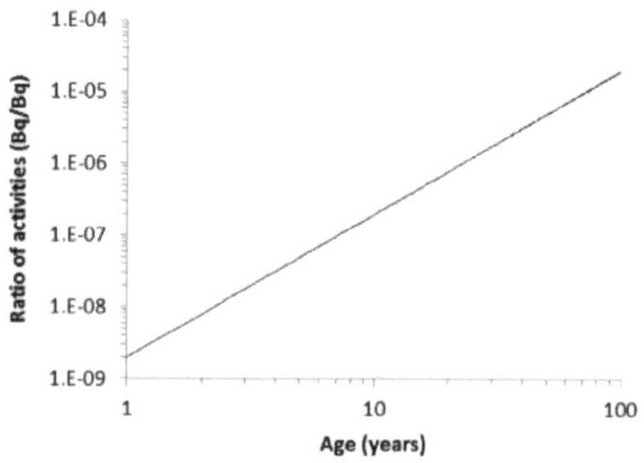

Figure 16.3: Ratio of ^{226}Ra/^{234}U (= ^{214}Bi/^{234}U) plotted versus age. This can be used to determine how long ago a sample was enriched based on the activities of ^{234}U and ^{226}Ra or ^{214}Bi in the sample.

Bibliography

G. Choppin, J.-O. Liljenzin, and J. Rydberg. *Radiochemistry and Nuclear Chemistry*, Elsevier Press, Oxford (2006).

W. Loveland, D. J. Morrissey, and G. T. Seaborg. *Modern Nuclear Chemistry*, Wiley, New Jersey (2006).

K. Mayer, M. Wallenius, and I. Ray, Nuclear forensics-a methodology providing clues on the origin of illicitly trafficked nuclear materials. *Analyst*, 2005, **130**, 433.

J. T. Mihalczo, J. A. Mullens, J. K. Mattingly, and T. E. Valentine, Physical description of nuclear materials identification system (NMIS) signatures. *Nucl. Instrum. Meth. A*, 2000, **450**, 531.

J. T. Mihalczo, J. K. Mattingly, J. S. Neal, and J. A. Mullens, NMIS plus gamma spectroscopy for attributes of HEU, PU and HE detection. *Nucl. Instrum. Meth. B*, 2004, **213**, 378.

K. J. Moody, I. A. Hutcheon, and P. M. Grant. *Nuclear Forensic Analysis*, CRC Press, Boca Raton (2005).

C. T. Nguyen, Age-dating of highly enriched uranium by γ-spectrometry. *Nucl. Instrum. Meth. B*, 2005, **229**, 103.

C. T. Nguyen, and J. Zsigrai, Basic characterization of highly enriched uranium by gamma spectrometry. *Nucl. Instrum. Meth. B*, 2006, **246**, 417.

D. Reilly, N. Ensslin, and H. Smith, Jr. *Passive Nondestructive Assay of Nuclear Materials*, National Technical Information Service, Springfield, VA (1991).

M. Wallenius, A. Morgenstern, C. Apostolidis, and K. Mayer, Determination of the age of highly enriched uranium. *Anal. Bioanal. Chem.*, 2002, **374**, 379.

176

Chapter 17: Refinement of Crystallographic Positional Metal Disorder in Molecular Solid Solutions

Simon G. Bott and Andrew R. Barron

Introduction

Crystallographic positional disorder is evident when a position in the lattice is occupied by two or more atoms; the average of which constitutes the bulk composition of the crystal. If a particular atom occupies a certain position in one unit-cell and another atom occupies the same position in other unit cells, the resulting electron density will be a weight average of the situation in all the unit cells throughout the crystal. Since the diffraction experiment involves the average of a very large number of unit cells (ca. 10^{18} in a crystal used for single crystal X-ray diffraction analysis), minor static displacements of atoms closely simulate the effects of vibrations on the scattering power of the average atom. Unfortunately, the determination of the average atom in a crystal may be complicated if positional disorder is encountered.

Crystal disorder involving groups such as CO, CN and Cl have been documented to create problems in assigning the correct structure through refinement procedures. However, disorder of elements at an individual atom site offers a greater challenge. There are two very different issues that must be considered when solving a crystal structure with site occupancy disorder:

- What is the relationship of a single crystal to the bulk material?
- Is the refinement of a site-occupancy-factor actually gives a realistic value for % occupancy when compared to the "actual" % composition for that particular single crystal?

The following represents a description of a series of methods for the refinement of a site occupancy disorder between two atoms (e.g., two metal atoms within a mixture of isostructural compounds).

Methods for X-ray diffraction determination of positional disorder in molecular solid solutions

An atom in a structure is defined by several parameters: the type of atom, the positional coordinates (x, y, z), the occupancy factor (how many atoms are at that position) and atomic displacement parameters (often called temperature or thermal parameters). The latter can be thought of as being a picture of the volume occupied by the atom over all the unit cells and can be isotropic (1 parameter defining a spherical volume) or anisotropic (6 parameters defining an ellipsoidal volume). For a normal atom, the occupancy factor is fixed as being equal to one, and the positions and displacement parameters are refined using least-squares methods to values in which the best agreement with the observed data is obtained. In crystals with site- disorder, one position is occupied by different atoms in different unit cells. This refinement requires a more complicated approach. Two broad methods may be used:

- either a new atom type that is the appropriate combination of the different atoms is defined,
- the same positional parameters are used for different atoms in the model, each of which has occupancy values less than one, and for which the sum is constrained to total one.

In both approaches, the relative occupancies of the two atoms are required. For the first approach, these occupancies have to be defined. For the second, the value can be refined. However, there is a relationship between the thermal parameter and the occupancy value so care must be taken when doing this. These issues can be addressed in several ways.

Method 1

The simplest assumption is that the crystal from which the X-ray structure is determined represents the bulk sample was crystallized. With this value, either a new atom type can be generated that is the appropriate combination of the measured atom type 1 (M) and atom type 2 (M') percent composition or two different atoms can be input with the occupancy factor set to reflect the percent composition of the bulk material. In either case the thermal parameters can be allowed to re ne as usual.

Method 2

The occupancy values for two atoms (M and M') are refined (such that their sum was equal to 1), while the two atoms are constrained to have the same displacement parameters.

Method 3

The occupancy values (such that their sum was equal to 1) and the displacement parameters are refined independently for the two atoms.

Method 4

Once the best values for occupancy is obtained using either Methods 2 or 3, these values were fixed, and the displacement parameters are allowed to refine freely.

A model system

Metal β-diketonate complexes (Figure 17.1) for metals in the same oxidation state are isostructural and often isomorphous. Thus, crystals obtained from co-crystallization of two or more metal β-diketonate complexes [e.g., $Al(acac)_3$ and $Cr(acac)_3$] may be thought of as a hybrid of the precursors; that is, the metal position in the crystal lattice may be defined as having the average metal composition. A series of solid solutions of $Al(acac)_3$ and $Cr(acac)_3$can be prepared for study by X-ray diffraction, by the crystallization from acetone solutions of specific mixtures of $Al(acac)_3$ and $Cr(acac)_3$ (Table 17.1, Column 1). The pure derivatives and the solid solution, $Al_{1-x}Cr_x(acac)_3$, crystallize in the monoclinic space group $P2_1/c$ with $Z = 4$.

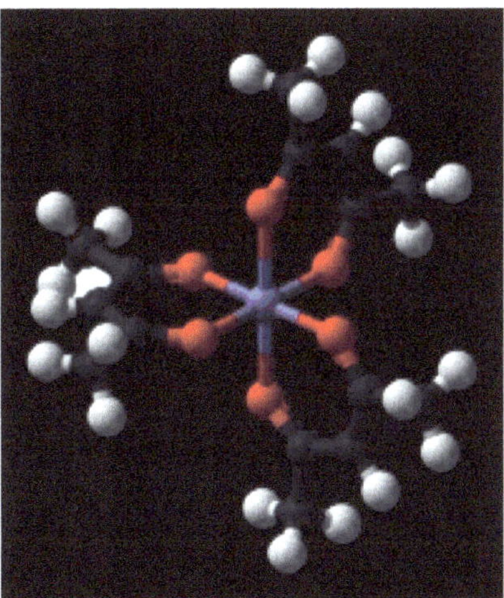

Figure 17.1: Molecular structure of M(acac)$_3$, a typical metal β-diketonate complex.

Substitution of Cr for Al in the $M(acac)_3$ structure could possibly occur in a random manner, i.e., a metal site has an equal probability of containing an aluminum or a chromium atom. Alternatively, if the chromium had preference for specific sites a super lattice structure of lower symmetry would be present.

Such an ordering is not observed since all the samples show no additional reflections other than those that may be indexed to the monoclinic cell. Therefore, it may be concluded that the $Al(acac)_3$ and $Cr(acac)_3$ do indeed form solid solutions: $Al_{1-x}Cr_x(acac)_3$.

Solution composition (% Cr)	WDS composition of single crystal (% Cr)	Composition as refined from X-ray diffraction (% Cr)
13	1.9 ±0.2	0[a]
2	2.1 ±0.3	0[a]
20	17.8 ±1.6	17.3 ±1.8
26	26.7 ±1.7	28.3 ±1.9
18	48.5 ±4.9	46.7 ±2.1
60	75.1 ±4.1	72.9 ±2.4
80	91.3 ±1.2	82.3 ±3.1

Table 17.1: Variance in chromium concentrations (%) for samples of the mixed metal $Al_{1-x}Cr_x(acac)_3$ crystallized from solutions of $Al(acac)_3$ and $Cr(acac)_3$. [a]Concentration too low to successfully refine the Cr occupancy.

Electron microprobe analysis, using wavelength-dispersive spectrometry (WDS), on the individual crystal from which X-ray crystallographic data provides the actual composition of each crystal. Analysis performed on at least 6 sites on each crystal using a 10 µm sized analysis spot providing a measure of the homogeneity within the individual crystal for which X-ray crystallographic data was collected. An example of a SEM image of one of the crystals and the point analyses is given in Figure 17.2.

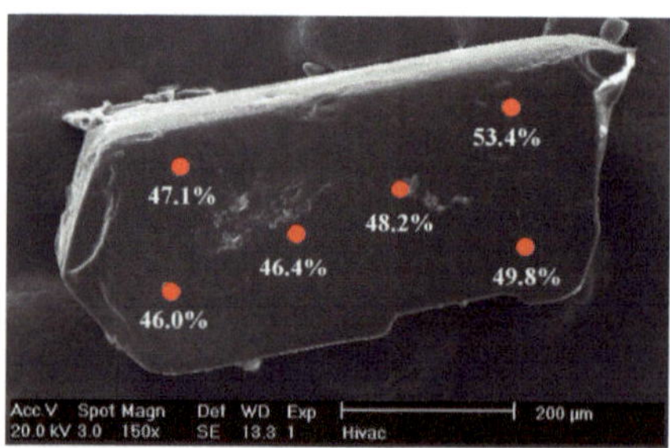

Figure 17.2: SEM image of a representative crystal used for WDS and X-ray diffraction analysis showing the location and results for the WDS analysis. The 10 µm sized analysis spots are represented by the red dots. Adapted from B. D. Fahlman, Ph.D. Thesis, Rice University, 2000.

The data in Table 17.1 and Figure 17.2 demonstrate that while a batch of crystals may contain individual crystals with different compositions, each individual crystal is actually reasonably homogenous. There is, for most samples, a significant variance between the molar Al:Cr ratio in the bulk material and an individual crystal chosen for X-ray diffraction. The variation in Al:Cr ratio within each individual crystal (±10%) is much less than that between crystals.

Comparison of the methods

Method 1

Since Method 1 does not refine the %Cr and relies on an input for the Al and Cr percent composition of the "bulk" material, i.e., the %Cr in the total mass of the material (Table 17.1, Column 1), as opposed to the analysis of the single crystal on which X-ray diffraction was performed (Table 17.1, Column 2) the closer these values were to the "actual" value determined by WDS for the crystal on which X-ray diffraction was performed (Table 17.1, Column 1 versus 2) then the closer the overall refinement of the structure to those of Methods 2 - 4. While this assumption is obviously invalid for many of the samples, it is one often used when bulk data (for example, from NMR) is available. However, as there is no reason to assume that one crystal is completely representative of the bulk sample, it is unwise to rely only on such data.

Method 2

This method always produced final, refined, occupancy values that were close to those obtained from WDS (Table 17.1). This approach assumes that the motion of the central metal atoms is identical. While this is obviously not strictly true as they are of different size, the results obtained herein imply that this is a reasonable approximation where simple connectivity data is required. For samples where the amount of one of the elements (i.e., Cr) is very low so low a good refinement cannot often be obtained. In these cases, when refining the occupancy values, that for Al would exceed 1 while that of Cr would be less than 1.

Method 3

In some cases, despite the interrelationship between the occupancy and the displacement parameters, convergence was obtained successfully. In these cases, the refined occupancies were both slightly closer to those observed from WDS than the occupancy values obtained using Method 2. However, for some samples with higher Cr content the refinement was unstable and would not converge. Whether this observation was due to the increased percentage of Cr or simply lower data quality is not certain. While this method does allow refinement of any differences in atomic motion between the two metals, it requires extremely high-quality data for this difference to be determined reliably.

Method 4

This approach adds little to the final results.

Correlation between analyzed composition and refined composition

Figure 17.3 shows the relationship between the chromium concentration (%Cr) determined from WDS and the refinement of X-ray diffraction data using Methods 2 or 3 (labeled in Figure 17.3).

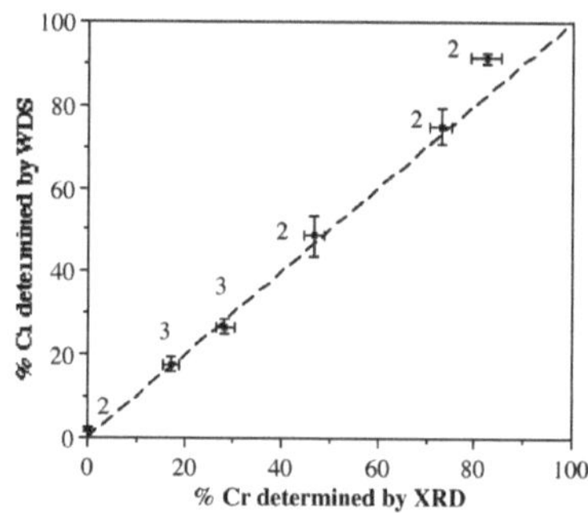

Figure 17.3: Comparison of the chromium concentration determined from WDS (with error) and refinement of X-ray diffraction data (with error) using Methods 2 or 3. Adapted from B. D. Fahlman, Ph.D. Thesis, Rice University, 2000.

Clearly there exists a good correlation, with only a slight divergence at high Cr concentration. This is undoubtedly a consequence of trying to refine a low fraction of a light atom (Al) in the presence of a large fraction of a heavier atom (Cr). X-ray diffraction is, therefore, an accurate method of determining the M:M' ratios in crystalline solid solution.

Bibliography

S. G. Bott, B. D. Fahlman, M. L. Pierson, and A. R. Barron, An accuracy assessment of the refinement of partial metal disorder in solid solutions of Al(acac)$_3$ and Cr(acac)$_3$. *J. Chem. Soc., Dalton Trans.*, 2001, 2148.

K. R. Dunbar and S. C. Haefner, Crystallographic disorder in the orthorhombic form of carbonyl(chlorobis(triphenylphosphine)rhodium: relevance to the reported structure of the paramagnetic impurity in Wilkinson's catalyst. *Inorg. Chem.*, 1992, **31**, 3676.

J. Glusker, M. Lewis, and M., Rossi, *Crystal Structure Analysis for Chemists and Biologists*, VCH, New York (1994).

K. Yoon and G. Parkin, Bond-stretch isomerism in the chlorooxomolybdenum complexes cis-mer-MoOCl$_2$(PR$_3$)$_3$: a reinvestigation. *Inorg. Chem.*, 1992, **114**, 31.

Chapter 18: Preparation of Mineral Samples for X-ray Fluorescence Spectroscopy

Johnathan Dietz and Andrew R. Barron

From sediment to sample

Sample sediments are typically sent in a large plastic bag inside a brown paper bag labeled with the company or organization name, drill site name and number, and the depth the sediment was taken (in meters).

The first step in determining a lithology is to prepare a sample from your bulk sediment. To do this, you will need to crush some of the bulk rocks of your sediment into finer grains (Figure 18.1). You will need a hard surface, a hammer or mallet, and your sediment. An improvised container such as the cardboard one shown in Figure 18.1 may be useful in containing fragments that try to escape the hard surface during vigorous hammering. Remove the plastic sediment bag from the brown mailer bag. Empty approximately 10-20 g of bulk sediment onto the hard surface. Repeatedly strike the larger rock sized portions of the sediment until the larger units are broken into grains that are approximately the size of a grain of rice.

Figure 18.1: A hammer and hard surface for crushing. The makeshift cardboard shield on the left can be placed around the hard surface to control fragmentation.

Some samples will give off oily or noxious odors when crushed. This is because of trapped hydrocarbons or sulfurous compounds and is normal. The next step in the process, washing, will take care of these impurities and the smell.

Once the sample has been appropriately crushed on the macro scale, a micro uniformity in grain size can be achieved through the use of a pulverizing micro mill machine such as the Planetary Mills Pulverisette 7, shown in Figure 18.2.

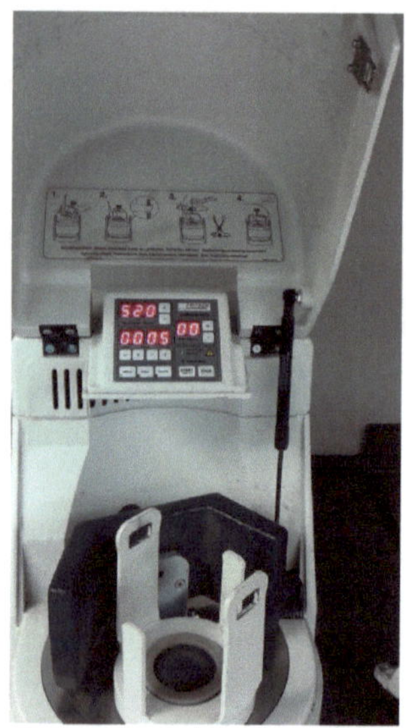

Figure 2.123: A Pullsette micro mill, milling cup removed. The mill is set to 520 rotations per minute and a five-minute run time.

To use the mill, load your crushed sample into the milling cup (Figure 18.3) along with milling stones of 15 mm diameter. Set your rotational speed and time using the machine interface. A speed of 500-600 rpm and mill time of 3-5 minutes is suggested. Using higher speeds or longer times can result in loss of sample as dust. Load the milling cup into the mill and press start; make sure to lower the mill hood. Once the mill has completed its cycle, retrieve the sample and dump it into a plastic cup labelled with the drill site name and depth in order to prepare it for washing. Be sure to wash and dry the mill cup and mill stones between samples if multiple samples are being tested.

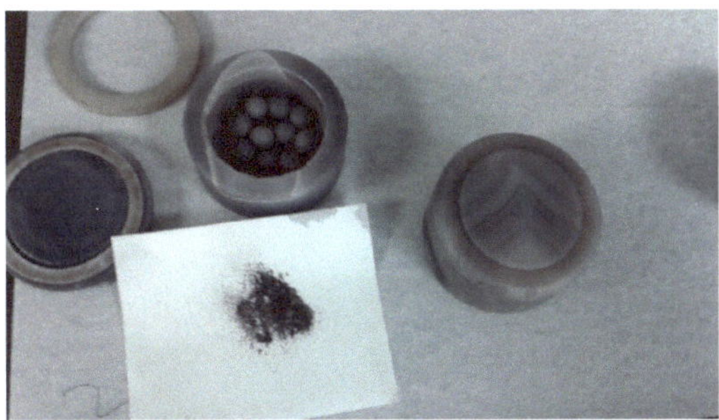

Figure 18.3: A milling cup with mill stones and the crushed sample before milling.

Washing the sample

If your sample is dirty, as in contaminated with hydrocarbons such as crude oil, it will need to be washed. To wash your sample, you will need your sample cup, a washbasin, a spoon, a 150-300 μm sieve, household dish detergent, and a porcelain ramekin if a drying oven is available (Figure 18.4).

Figure 18.4: A washbasin, with detergent in a squirt bottle and the sample in a cup for washing (sieve not pictured).

Take your sample cup to the wash basin and fill the cup halfway with water, adding a squirt of dish detergent. Vigorously stir the cup with the spoon for 20 seconds, ensuring each grain is coated with the detergent water. Pour your sample into the sieve and turn on the faucet. Run water over the sample to allow the detergent and dust particles to wash through the sieve. Continue to wash the sample this way until all the detergent is washed from the sample. Once clean, empty the sieve onto a surface to leave to dry overnight, or into a ramekin if a drying oven is available. Place ramekin into drying oven set to at least 100 $^\circ$C for a minimum of 2 hours to allow thorough drying (Figure 18.5). Once dry, the sample is ready to be picked.

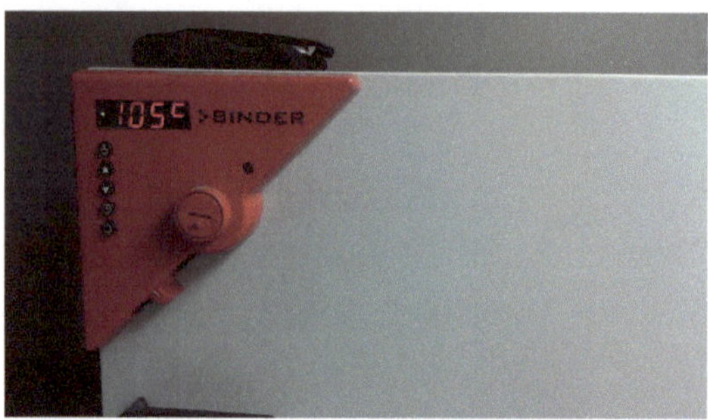

Figure 18.5: A drying oven with temperature set above the temperature of evaporation for water at 105 °C.

Picking the sample

Picking the sample is arguably the most important step in determining the lithology (Figure 18.6). During this step you will create a sample uniformity to eliminate random minerals, macro contaminates such as wood, and drop-stones that dropped into your sediment depth when the sediment was drilled. You will also be able to get a general judgment as to the lithology after picking, though further analysis is needed if chemical composition is desired. Remove sample from drying oven. Take a piece of weighing paper and weigh out 5-10 g of sample. Use a light microscope to determine whether most of the sample is either silt, clay, silty-clay, or sand.

- Clay grains will have a gray coloration with large at sub-surfaces and less angulation. Clay will easily deform under pressure from forceps.
- Silt grains will be darker than clay and will have specks that shine when the grain is rotated. Texture is long pieces with jagged edges. Silt is harder in consistency.
- Silty clay is a heterogenous mixture (half and half mixture) of the above.
- Sand is defined as larger grain size, lighter and varied coloration, and many crystalline substructures. Sand is hard to deform with the forceps.

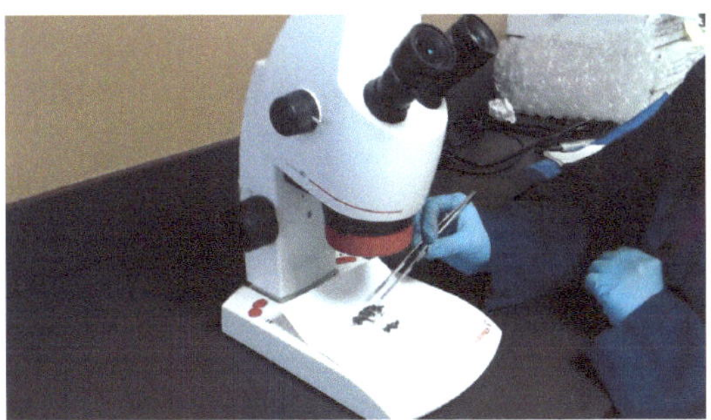

Figure 18.6: A light microscope being used to 'pick' the sample. The sample is being separated according to the dominant lithology in preparation for chemical analysis.

Pelleting the sample

To prepare your sample for X-ray fluorescence (XRF) analysis you will need to prepare a sample pellet. To pellet your sample, you will need a mortar and pestle, pellet binder such as Cerox, a scapula to remove binder, a micro scale, a pellet press with housing, and a pellet tin cup. Measure out and pour 2-4 g of sample into your mortar. Measure out and add 50% of your sample weight of pellet binder. For example, if your sample weight was 2 g, add 1 g of binder. Grind the sample into a fine, uniform powder, ensuring that all of the binder is thoroughly mixed with the sample (Figure 18.7).

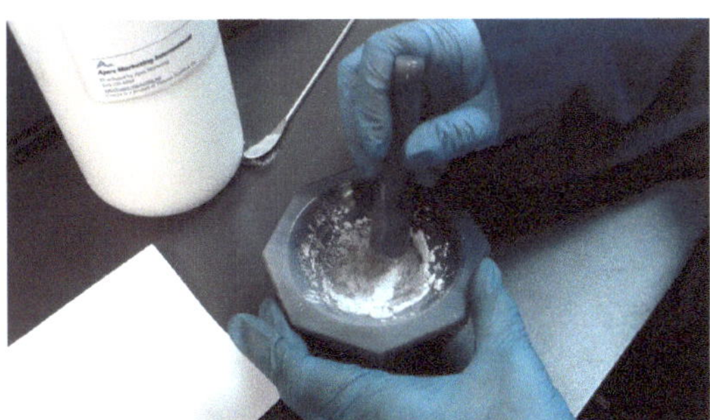

Figure 18.7: A mortar and pestle being used to grind the sample to a powder for pelleting in the pellet press. A binding agent (Cereox) is also added using the scapula.

Drop a sample of tin foil into the press housing. Pour sample into the tin foil, and then gently tap the housing against a hard surface two to three times to ensure sample settles into the tin. Place the top press disk into the channel. Place the press housing into the press, oriented directly under the pressing arm. Crank the lever on the press until the pressure gauge reads 15 tons (Figure 18.8). Wait for one minute, then twist the pressure release valve and remove the press housing from the press. Reverse the press and apply the removal cap to the bottom of the press. Place the housing into the press bottom side up and manually apply pressure by turning the crank on top of the press until the sample pops out of the housing. Retrieve the pelleted sample (Figure 18.9). The pelleted sample is now ready for X-ray fluorescence analysis.

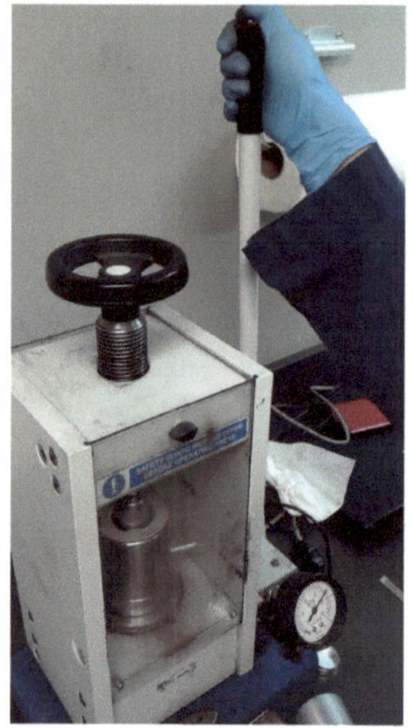

Figure 18.8: A pellet press being pressurized to 15 tons.

Figure 18.9: A completed pellet after pressing.

XRF analysis

Place the sample pellet into the XRF (Figure 18.10 and 18.11) and close the XRF hood. The XRF obtain the spectrum from the associated computer.

Figure 18.10: A Spectro XEPOS X-Ray fluorescence spectrometer.

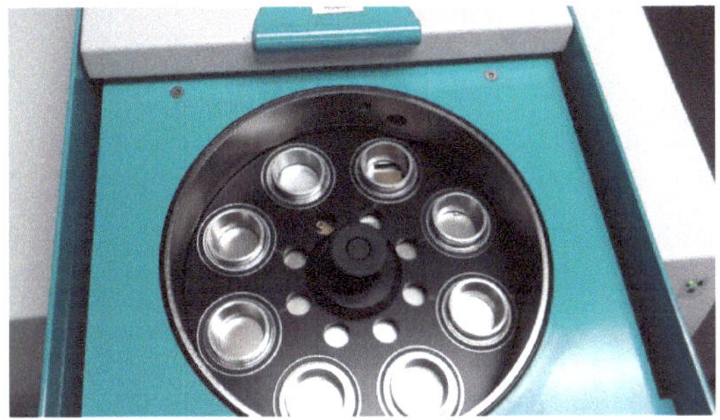

Figure 18.11: The inside of the spectrometer where the sample pellets are placed for analysis.

The XRF spectrum is a plot of energy and intensity. The software equipped with the XRF will be pre- programmed to recognize the characteristic energies associated with the X-ray emissions of the elements. The XRF functions by shooting a beam of high energy photons that are absorbed by the atoms of the sample. The inner shell electrons of sample atoms are ejected. This leaves the atom in an excited state, with a vacancy in the inner shell. Outer shell electrons then fall into the vacancy, emitting photons with energy equal to the energy difference between these two energy levels. Each element has a unique set of energy levels; therefore, each element emits a pattern of X-rays characteristic of that element. The intensity of these characteristic X-rays increases with the concentration of the corresponding element leading to higher counts and higher peaks on the spectrum (Figure 18.12).

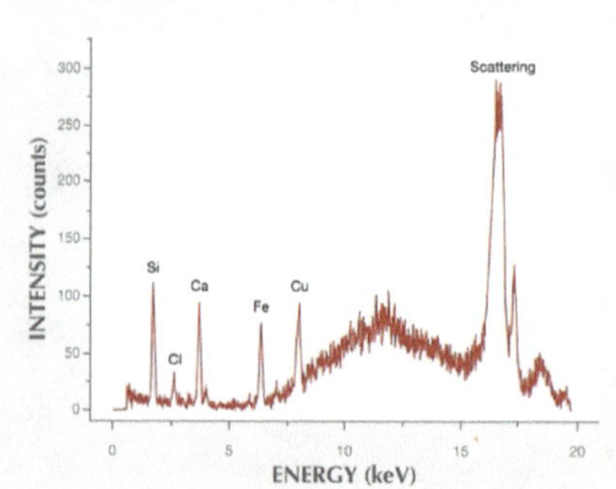

Figure 18.12: The XRF spectrum showing the chemical composition of the sample.

Bibliography

E. H. Williams, A manual of Lithology: *Treating of the Principles of the Science with a Special Reference to Megascopic Analysis*, Inman Press (2008).